# BETWEEN TWO MIRRORS

## MIRRORS

Art and science in our modern world

Gary McCallister

DEDICATION

TO DAD

## ACKNOWLEDGMENTS
Without Gaydra there would be no book and no children.
Without my children there would be no life.
Without life there would be no broken swords.

# CONTENTS

ACKNOWLEDGMENTS ............................... 1
1. BETWEEN TWO MIRRORS..................... 6
2. INTRODUCTION .............................. 9
3. A CROWNLESS KING .......................... 14
4. ALONE ...................................... 18
5. BACK FROM TOMORROW .................... 21
6. BALTHAZARS FEATS ......................... 25
7. BEGINNING UNDERSTANDING ............. 28
8. BEYOND THE VEIL ............................ 31
9. BYZANTIUM .................................. 35
10. CEMETARY PASSAGES ......................... 39
11. COME WALK A LONELY TRAIL WITH ME .42
12. FALLEN ANGEL ................................. 45
13. GUIDED BY THE CONSTELLATIONS ........50
14.. IN MY DREAMS .................................54
15. NO PLACE LEFT TO GO ........................57
16. SEE HERE MY BROKEN BLADE ..............61
17. SEEKER ALL ...................................65
18. TEARDOPS TURNED TO RHYME ............ 69
19. THE BUSINESS PLAN ........................... 73
20. THE DOCK .....................................77
21. THE GUIDE .................................... 80
22. MORTAL GARDEN ..............................84
23. THE PREACHER ................................. 88
24. THE TEACHER ..................................92
25.. THROUGH THE NARROWS ....................97
26. THINDER ON THE MOUNTAIN ..............100
27. TIME ..........................................104
28. THE PAINTER AND THE PAINTING ........107
29. UNITED........................................111
30. WANDERERS IN A STRANGE LAND .........114
31. SEVEN BLACK CROWS ..........................116
32. VISIONS .......................................119
33. DESERT DREAM ...............................122
34. CHAKO CANYON ...............................126
35. OUT OF CHAOS ...............................129
36. SEED...........................................132
37. SISTERS OF THE WOLFEN MOON............135
38. FLAMING MOTH................................138
39. DUST...........................................142
40. KINGS OLD.....................................145
ABOUT:
    THE AUTHOR...................................148
    MACDONALD ...................................148
OTHER BOOKS BY GARY MACCALLISTER........149

-1-

# BETWEEN TWO MIRRORS

*"The best thing you can do for your fellow, next
to rousing his conscience, is — not to give him things
to think about, but to wake things up that are in him;
or say, to make him think things for himself."*
*George MacDonald*

This book is concerned with the arts, the sciences and the two together. These two subjects are often seen as opposites, and I suppose they are in some respects. Yet the two subjects also share much in common.

SCIENCE

Science is concerned with physical things and such questions as: How many are there? How does a falling thing fall? What shape is it? How big is it. How much does it weigh? Why does that object act as it does? The questions require scientists to restrict their attention to a single object or event, and study that one object or event carefully. Science only works on the material world.

Scientific study requires physical skills learned through the acquisition of special techniques. A scientist may have to invent new methods and perfect skills to be able to conduct his studies. Often studies are done to establish a pattern or direction. But from this careful and sometimes lengthy process, the scientist attempts to distil some kind of general understanding about the object or event that they have studied. This general understanding is sometimes called a theory.

As the theory becomes more reliable and useful, it can sometimes become a law. These general ideas can then be used to compare other similar objects, evaluate a theory further, and make predictions about events under certain conditions.

But, overall, it appears that scientists begin with some real-world physical object or phenomenon and conclude with a general idea called a hypothesis or theory. They turn the world of reality into a world of imagination, thought, and abstraction.

ART

By contrast, art is first concerned with ideas. Much of art, including visual art, music, language arts and the performing arts, appears to be born from such thought as: religious concepts, political movements, cultural characteristics, imaginary events, or social ideals. These require the artists to restrict their attention and focus on a specific idea they wish to explore.

Art also requires the use of physical skills and special techniques. The artist may have to invent new methods and further refine his skills to be able to conduct his study. Often the artist must make several models, or be involved in many attempts, before capturing his ideas into a tangible form.

In the end, the artist creates a physical object which represents his view of an idea. The end product of art is a function of the physical world. It may be visual, audible, or even palpable; but it is real. This object can then be used to test the accuracy of the artists', and societies, understanding of the idea, explore the ramifications of the idea, explain the idea more fully to others, or even test the truthfulness of the idea.

But the overall conclusion to be drawn is that artists tend to begin with some non-physical idea and conclude with a real object or physical manifestation that can be detected by the human senses. Artists turn the imaginary and abstract world of ideas into material objects and reality.

ART AND SCIENCE

It would seem that both scientist and artist are concerned with understanding our world while arriving at some form of truth and increased understanding. Both utilize existing knowledge, personal skill and equipment. What appears significantly different is that they initiate their mental journeys from separate, even opposite, starting points.

Because of their opposite trajectories, scientists and artists often see themselves as "in conflict". But understanding the similarities of the two endeavors enriches each field significantly. Recognizing this can be especially powerful in educational endeavors. Numerous studies and pilot projects have shown that using one approach to study the other is especially effective.

For example, having students write about math or science has increased understanding of these fields for many students. Writing computer programs that artistically animate scientific phenomenon has proven animation to be an excellent teaching and learning tool. Assigning an artist to explore a specific scientific concept in an art class leads to greater understanding of both art and science.

Perhaps the world needs fewer engineers, scientists, poets, and musicians. What is needed is more people who understand the relationship between ideas and objects. The creation of ideas has an effect on the objects. The creation of objects has an effect on the creation of ideas.

**BETWEEN TWO MIRRORS**

"Hey, I haven't seen you forever.
How've you been?"

> "Can't complain. How about you?
> You're looking thin."

"Nice of you to notice.
But who's that behind you?"

> "That's just me like I used to be.
> I thought you knew.
> But did I interrupt?
> Who was that you were talking to?"

"Oh, that's the guy I'll be
When I'm finally through.
But look. That guy looks familiar
Whose coming this way."

> "Well, of course.
> That's you looking back at today."

"Look at this crowd,
People as far as I can see."

> Yeah, it's just the same behind you.
> I completely agree."

"Seems this coming and going never ends."
> "Hey, it's been good to see you, my friend."

-2-

# STRANGE TALES

*"I write, not for children, but for the child-like,*
*whether they be of five, or fifty, or seventy-five."*
George MacDonald

Once upon a time, there was no time. I suppose you are expecting a fairy tale. However, everyone knows that fairy tales aren't real. Except they are in a strange way. Fairy tales are sometimes even more real than true stories. Science tells true stories. Everyone also knows that science stories are sometimes more strange than fairy tales. And science says that, "Once upon a time there was no time."

PRE-TIME:
What existed prior to time? That was eternity, or pre-time. Eternity isn't time.

It's like a novel. Events in a novel take place in the time of the novel. However, the time in the novel has no meaning in the author's time other than the time it takes the author to write about time in the novel.

So, events in pre-time take place in their own time which has nothing to do with our time. Look, it's not all that different from Daylight Savings Time where we don't save any time at all!

Consider the events in a novel. If I write a novel in which a murder is committed in novel-time, it doesn't apply to me. I cannot be arrested for a "novel murder" unless it is a murder in real time committed in a novel way. I am not part of the events of a novel, and my time is not part of the novel's time. I am eternal from the point of view of a novel. That's why stories always occur in "once upon a time".

PAST:
However, scientists today do believe that, at one time, there was no time. This is because "time" is a dimension in which events can be ordered in sequence from the past, to the present, to the future. Events, on the other hand, are descriptions of things that happen in a material world, that can be described by physical laws, and that can be ordered in time. So if there was ever a time when there were no physical happenings, there would be a time when there was no time.

Modern physicists think that there was a time when there was nothing happening.  The first event was a massive explosion which created the universe, sort of like, "Let there be light."  All the matter in the universe came into existence at that time as well as all the laws of physics.  Since there were no events prior to that time, there was no time.  Time came into existence when things started to happen that could be sequenced.

Thomas Aquinas, who lived more than 800 years ago, maintained that every physical event has a cause.  He further pointed out that the cause of any physical event had a previous cause, and that previous cause would have also had a previous cause.  If there was ever a time when there were no previous causes, there would also be no time.

PRESENT:

Sir Isaac Newton, on the other hand, had a so-called realistic view of time.  He saw it as a container in which things happened in sequence.  So let's talk about time in volumetric ways like one would a container.  "I'd like an extra quart of time, please."  "Yes, I believe I'll have two scoops of time."  That doesn't work too well, does it?

Immanuel Kant thought time was a purely intellectual concept, like numbers, and couldn't be traveled or measured.  Dr. Who fans will find this idea unacceptable.  Of course, that doesn't ring true either because there seems to be a distinct difference between the time the alarm rings and the time I actually get up.  These are usually two significantly different events.

"Having not enough time" is not the same as "having no time".  Not having enough time is incorrect.  Everyone has all the time there is available at the time.  It isn't that we don't have enough time.  It's that we have too many events occurring in a given time frame.  If we cut out some events, we find that we had more time, even though the amount of time we are talking about is the same as before, when we had more events.

FUTURE:

I suppose the same reasoning can take place about post-time which, unfortunately, we also call eternity.  When all events stop happening in our time, then we are no longer in time.  If we used to be in time, but are no longer, a condition we sometimes call "dead", we are in post-time.  I suppose the events in post-time bear little resemblance to events in time just as events in pre-time are not the same as events in time.

In fact, it's again like in a novel. The characters in post-time may have experiences, but they are not happening in time. If I write a novel set in post-time, I cannot be held accountable for the things that happen in post-time because I am in time as I write and the events are outside of time.

I think this is why so many people like to write novels. Authors get to play at creating worlds, setting them in motion, and watching them play out in "novel-time". They can't be held responsible for what goes on there because they are living outside of that experience or time.

NOW:

Unfortunately, we are caught in between the past and the present. It's hard to know whether or not that is unfortunate, or fortunate, because none of us know or understand what we were in pre-time or exactly what it will be like in post-time. However, I prefer to safely assume that I am here now!

Of course, like Joshua speaking to the Children of Israel before they entered the promised land, we have all been given "A land for which ye did not labor, and cities which ye built not, and ye dwell in them; of the vineyards and oliveyards which ye planted not do ye eat."

Each generation exercises power over their ancestors and their posterities. A generation can be grateful for what they are given, or they can rebel against their traditions and destroy what is built. This resists and limits the power of their predecessors. Or, they can set about modifying the environment they inherit by bequeathing a different world to their offspring.

In this way, we are as one standing before facing mirrors. We see the future and the past ever more dimly in the mirrors before and behind us. We are left to contemplate where we have come from, why we are here, and where we are going.

The following collection of essays and poems are fairy tales, and science tales of some kind. Some explore todays present. Some explore the present from a time before we are born. Some tell tales about the present told from after the participants are dead. Some take place after the participant is dead. Some explore the transition between the two borders. For others, it's simply hard to tell.

## STRANGE TALES

Black bird fluttered 'round the carcass,
Strutting o'er the feast it found.
Then it looked and when it saw me,
Slowly lifted off the ground.
"Follow me!" it seemed to say,
Flapped its wings and flew away.
'Follow me, across the hills
Into the canyon cold and still."
There it perched on a thorny limb
And croaked a solitary hymn.
It looked around to be sure
That I had followed him.
To me it was crystal clear.
The bird was saying, "Over here."
So I slowed from my run
Just to see what the bird had done.
Slower now, but sweaty still,
In canyon shadow, I felt a chill.
As I approached the broken tree,
The bird screamed both loud and shrill.
Then the bird flew away,
Soaring up into the sky.
It tore a hole in the blue
And disappeared it flew so high.
It left behind a feather there,
Black and long on the earth so bare.
It fluttered down and laid so still.
What had been written by that quill?

In the sand beneath the tree,
I don't know when last eyes had seen,
Stood a spade rusted and old,
And I thought "I've found my gold."
So I began to dig in the earth,
Sure to find my treasure there.
But my spade stopped in midair.
I saw a skull with long black hair.
My heart paused with a stab of fear.
Who could it be buried here?

My mind said for me to run.
But a darker side said, "You're not done."

I found two bodies buried there.
They were lovers I would guess.
Lay entwined in each other's arms,
But she had a knife within her chest.
He held her close, so tenderly.
But, when unearthed, I could see.
Gun in his hand on his funeral bed.
And a bullet hole in his head.

I dropped my spade and began to run,
Frightened by a sudden pain.
For I thought I knew this pair,
Though the lonely grave bore no names.
For my love had long black hair,
And I had stabbed her without a prayer.
But if it was me whose life I gave,
Who had buried us in that grave?

# 3

# A CROWNLESS KING

*There are thousands willing to do great things*
*for one willing to do a small thing.*
George MacDonald

I've tried to explain to my wife that it isn't as easy to be a professional windbag as it looks. I spent years in preparation: undergraduate degree, graduate school, scientific conferences, and an unbelievable amount of practice shooting bull in the military. Somehow, she thinks what I do isn't work.

I think she has been confused by modern physics. According to physicists, work is equal to force times distance (W=F x D). Like me, you probably remember that from high school. It didn't make sense then either. You mean, if nothing moves, then no work is being done? This is where the average person begins to lose faith in science. I have spent years building powerful muscles that enable me to sit for long periods of time without tiring, and now you're telling me it was all for naught.

Since I sit, unmoving, at a computer most of the day writing words or music, she thinks that nothing is moving. No force is being applied, and so no work is being done. See how misleading physicists can be?! This is one reason physicists are never elected president or anything. You can't trust them to make sense.

What are all these mysterious forces anyway? Have you ever seen one? They tell me forces are measured in Newtons. But Newtons are simply an abstract idea arrived at by multiplying mass times distance divided by the time squared. You probably remember that from High School also.

$$N = \frac{M \times D}{T^2}$$

Anyway, in an effort to justify calling what I do work, I decided to investigate the process using physics, itself, to defend my activity. My hypothesis is that if computer keys require a force to press them down over a distance, then work is involved.

The average computer keyboard weighs 900 grams (gms). Since there are 106 keys on my keyboard, I can estimate that each key weighs about eight grams. Of course, that is an overestimate since keys are just the façade over the working mechanisms and they're contained within the surrounding frame. Therefore, I am going to estimate that each key only weighs about five grams or 0.005 kg. I actually think it is less than that, but I am fudging the data to support my hypothesis.

When a computer key is depressed, it moves on average, two millimeters (mm) or 0.002 meters. Since force is applied over time, I estimate that a key stroke takes about 0.2 seconds (sec.). So I now have mass, distance and time with which to calculate the amount of work in one key stroke.

$$N = M \times D \quad \frac{0.005 \text{ kg} \times 0.002 \text{ m}}{0.04 \text{ sec}} = 0.00025 \text{ Newtons}$$

$$W = F \times D \quad 0.00025 \times 0.002 \text{ m} = 0.0000005 \text{ Newton-meters (= joule)}$$

I admit that 0.0000005 Newton-meters doesn't sound like much. However, in a 650-word column, there are approximately seventy keystrokes per line and on average sixty lines. That is 3600 keys strokes which would be 0.018 Joules. Now that is a low estimate because I only use two fingers to type, so I have to move my hands a lot more than most. I figure that at least doubles the amount of work to 0.0036 joules for one column alone. That sounds like work to me!

Considering that I do this day in and day out for hours on end, I figure I do plenty of work. For example, in my most recent book, "A Convenient Truce: a cease fire between religion and science", available now from Amazon in both paperback and kindle formats, has 60,000 words. This is approximately 180,000 key strokes, or 0.09 joules. Ha!

Being a professional windbag is just as difficult as any other activity if you're really going to do it well. Some mass still has to move over some distance within a given time frame that is squared. What is squared time, by the way? I may make it look easy, but that doesn't mean that it isn't all just hot air after all.

## A CROWNLESS KING

Nothing to lose because it's already lost.
I stumble awake, it's time to begin.
The day approaches of the Pentecost
And I turn away from what might have been.

> I pitched the hay in the sun
> From early morning until the day was done.
> With flowing motion, the forks graceful arc,
> Dancing to the song of the meadowlark.

A fool cannot make another wise.
The future not better than the time before.
Those who say so are full of lies.
Each child must learn what is in store.

　　With shoulders broad and mighty arms,
　　I knew the joy of the farm.
　　Golden hay sailed through the air;
　　Golden piles like golden hair.

Growth without, without growth within,
Makes another page of history.
Our lives determined by what has been
Though we determine what will be.

　　A nobleman, who has labored not,
　　Watched in wonder at what I wrought.
　　He offered gold if I would come
　　And show the court how my dance is done.

Inside I burn, but outside I'm cold.
I stand alone in a crowded room.
My labor now is sold for gold,
And I fear the rose has lost its bloom

　　Each day I pitched imagined hay
　　For just one hour of the day.
　　Until only a simple week had passed,
　　And emptiness is all I had amassed.

Men should not go where I have been.
Men should have something to give.
Doing nothing is a sin.
Doing something is why we live.

　　So I resigned my effortless task.
　　"But why?" the noble master asked.
　　Do I not pay sufficient wage
　　To perform upon my gilded stage?

I can see now what life can give,
Though it seems against all reason.
If a man is to truly live,
He must labor for his seasons.

  Master thou art truly kind.
  You may think I've lost my mind.
  Though I may be a crownless king,
  I'll not do just anything.

Can you see beyond the tomb?
Can you see when you are gone?
You did not leave the womb
To be nothing but a pawn.

# 4

# ALONE

*"Only he knew that to be left alone is
not always to be forsaken."*
*George MacDonald*

I no longer have a mobile device. Apparently this is shocking, dangerous, and anti-social behavior. Some people are appalled, although most of them kindly ignore it like they would a birth defect. Others are openly irritated that I am so selfish as to be difficult to get ahold of. Some people  think it is dangerous to not be in touch with other people.

I'm not opposed to technology. I use it a lot to make music and to write. This isn't a protest or a moral issue. It simply got depressing for me to have a cell phone that never rang while I was surrounded by people who were always talking on theirs. It drew attention to the obvious. I have no friends. It's much easier to be above the fray than to have no fray. Anyway, I gave my phone to someone with friends.

Being connected is not the same as having friends, just like lonely is not the same as being alone. That's what I say anyway.

The word "alone" comes from Middle English and means literally "all one". So without an electronic appendage, I am one, complete, whole unit; alone. Bing alone used to be a common condition until technology made it obsolete.

On the other hand, the word "lonely" means 'being cut off from others". So obviously, if I have no phone, I am not connected to others. Therefore, I am incomplete and lonely. I am hoping my obvious loneliness will attract enough friends that I have to get another cell phone.

Many people seem to fear being alone. Maybe it is a fear of loneliness and the assumption that, if you are alone, you will be lonely. However, being alone can be pleasant, I'm told. In fact, it can be addictive. There used to be a twelve step program for "alone-addiction", but technology replaced it with cell phones. Now we need a twelve-step program to get over our phone addictions.

It seems like people are seldom alone. With family, school, extracurricular activities, and work there is no time left. However, if you should catch yourself being alone, you can always text or talk to entertain yourself. After all, your need to not be lonely is more important than someone else's need to be alone.

Being alone isn't all great, of course. There are times, when I am alone, that I get quite irritated with myself. I didn't understand this until I had children. When they started acting a lot like me, I realized how irritating I was! But I do get irritated with people who call when I want to be alone, so it's probably a wash.

Technology has so confused the issue of being alone and loneliness. Does Twitter reduce loneliness? Are you really alone if you are texting someone? If you are using Snapchat while you are by yourself, are you alone or lonely? For that matter, if you are talking on the phone, but there is no one present, what are you? These are good, solid, scientific questions that someone ought to look into. I'd do it myself, except I would have to get a cell phone to do the survey.

What if I called a person, when conducting my survey, to see if they want to be alone when they are trying to be alone? What do you think the response would be? How would I tell someone who wants to be alone from someone who is already with someone so doesn't want to take the survey right now? Psychological research can really be tricky.

I think people who like to hike, fish, birdwatch, hunt, and camp probably just want to be alone. People who don't do any of these things always remind you to be sure to take your cell phone along. It's a safety thing, they say. It makes me wonder how Columbus ever discovered America or the pioneers crossed the plains. Dangerous stuff, this going about without a cell phone.

**ALONE**
As I went out in the forest green
And saw white moths on nightly-wings,
I crowned myself King on a splintered throne
And sat in the middle of a fairy ring.
On the hunt, I ordered the hounds
And the noble stag bounded away.
Thus began the chase to the ground,
But first I bent my knee to pray.

Something rustled the grass not far away.
I thought perhaps I heard my name.
Did I imagine it? I cannot say.
But it came again just the same.

The sound came from a glimmering girl,
And I suddenly knew what I had not known.
She disappeared with a ghostly whirl,
And I became aware I was alone.

I started after her through the leaves,
But could not tell where she had gone.
I scraped my boot and tore my sleeve,
And through the forest I stumbled on.
The stag forgotten in my haste;
The stars, the wind, the fairy ring.
The useless throne seemed but disgrace.
One needs a queen to be a king.

Though now old, I am wandering still
Through the promised Holy land.
I found her on a distant hill,
Pressed her lips and held her hand.
We walk through dappled shadows
And soft whispering of the grass.
From time to time we pass to and fro
Until eternity is past.

# 5

# BACK FROM TOMORROW

*"If we will but let our God and Father work
His will with us, there can be no limit to
His enlargement of our existence"*
—*George MacDonald*

I guess I have always had a tenuous grasp on the concept of time. That may be overstating it a bit. But "Does anyone really know what time it is? Does anyone really care?" No, wait a minute. That was Chicago. Sometimes I think in lyrics. But that's only when I think, so it doesn't happen often.

Anyway, it just seems strange that I can remember the past, but I can't remember the future. Time and memory both take place in my brain, so what's up with that? Like last "Summertime, when the living was easy. . ." Where did that go? Yet, sometimes I think newscasts are never going to end.

I think I may have figured out my problem, though. I'm living in the past, and so are you. See, according to David Eagleman, there is an 80 millisecond delay between the time something happens and the time it is perceived. A millisecond is a thousandth of a second. Sure 80 milliseconds isn't much, but "As time goes by. . ." that is how much is delayed if you slice it thinly enough. By the time you are aware of anything, it has already happened.

It takes time for data to reach us. It must be translated into electrical impulses by our neural receptors and transmitted to the brain electronically. What really concerns me, though, is that it appears a lot of data either never reaches me, or I fail to turn it into electrical impulses, or it never gets transmitted. Hence, my tenuous grasp, I guess. It's hard to know where the breakdown occurs.

You can check out this whole idea yourself if you really want to. Go outside, have someone stand thirty meters away, and ask him to clap his hands. You will hear the clap at the same time you see it. Now have him move thirty one meters or more away to clap, and you will see the clap before you hear it. Just one stride will make the difference. You can occasionally see this phenomenon when the sound and sight on your television get out of synch by more than 80 milliseconds.

This 80-millisecond delay is kind of disturbing. If you think about it,

"The times they are a changing".  If what I perceive happened 80 milliseconds ago, and if what happened 80 milliseconds ago occurred in response to some other event that happened 80 milliseconds before that, then I am falling behind at the rate of 80 milliseconds per event. Now, like the song "Seven and a half cents", from the musical 'Pajama Game', eighty milliseconds doesn't mean a heck of a lot.  But at the rate of every perceived event, it can stack up pretty quickly.

Eighty milliseconds divides into one thousand milliseconds more than twelve times.  Just to be conservative, I am rounding down.  That means that I lose a second in my perceived life every twelve events.  I have no idea how many separate events I experience per second; but every twelve events, I am behind by an entire second.

I am not even sure what constitutes an event.  But it isn't hard to imagine that I am just now reacting to events that actually took place years ago! This thought has the potential to totally destroy much of Newtonian physics as well as the whole idea of cause and effect.  If every effect is collectively delayed from its cause by eighty milliseconds, we are in big trouble.

"If I could save time in a bottle", I might be able to catch up from time to time by dispersing some libation of time.  However, this does offer an explanation for why if "When the clock strikes two, three and four, the band slows down."  It's because they have been getting about eighty milliseconds behind the beat all night long.  By that time of night, everyone's grasp of time is a little tenuous.

Humans tend to be convinced that tomorrow will be better than today, except when they fear it will be worse.  I think it is partly generational.  Young people seem to think tomorrow will be better, partly because they think they know what to do to make it better. Older folks sometimes know, all too well, what can go wrong and may tend to decry and fear the future.

Sometimes, when things seem complicated, humans tend to long for some mythical past when everything was wonderful.  A couple of generations ago, regardless of when you live, almost always seems idyllic until you see photographs from the era and realize maybe it was sort of hard and dirty.

The truth seems to be that the future is the future.  It is constantly created by people who have not experienced the past.  My children did not experience my world, and my children's children will live in a different world than their parents did for most of their lives.  Each generation begins from a different beginning and, therefore, sees a different solution and goal.

The entire idea of progress implies a constant goal, an impossibility given the nature of change and generations. The future may look to the past for wisdom, and the present may set goals for the future. But it is a dog chasing its tail. In the end, both past and future are beyond our control.

## BACK FROM TOMORROW

I came back from tomorrow to see what we left behind.
We have gone too far to end it in a day.
All our efforts have left us as we started,
Now we're lost and cannot find our way.
In the future it all seems wrong.
It seems we started out so strong.
So I have returned to history
To look around and see what I can see.

Tomorrow claims to know the answers that have been found.
To be allowed to return is a mighty prize.
I had to convince the council and the crown
That past fools could make tomorrow wise.
Even then I had to fear
The monsters that guard the way to here.
Between us a void, cold and black,
That guards the path to the way back.

Tomorrow the multitude is searching in vain,
As if the answer could be found in men.
Because nothing is better than it used to be,
As if then is more wise than when.
Tomorrow's greatest mystery
Is after death what will be.
And like unto it, why we live?
Do we live to consume or to give?

My journey from tomorrow has been a blessed gift.
The lies that lie before the future are plain to see.
The truth that grows in each new man
Does not progress into history.
To see the joy, we were meant to know

Each generation's gift to grow.
To grow without is not to grow within.
Each generation commits its own sin.

# 6

# BALTHAZARS FEAST

*"Man finds it hard to get what he wants, because*
*he does not want the best; God finds it hard to give,*
*because He would give the best, and man will not take it."*
*George MacDonald*

My wife tends to keep boxes. She says that you never know when you might need one. We always have lots of them to choose from. They are out in the garage where the truck would be parked if we didn't have so many empty boxes.

Boxes are uniquely human, you know. They are akin to walking upright or opposable thumbs. There are a lot of people who want to think that humans are just another form of animal life, but I don't know of another species of animal that makes or uses boxes, let alone stores empty ones for later. It has got to be an advanced and unique characteristic demonstrating clearly that humans are something special. At least, my wife is.

There are some animals that store stuff, but they just use cavities in trees and holes in the ground. I suppose that helps them cut down on clutter, so they can park their trucks in the garage. I'm not sure if animals can remember where they store their stuff. Humans use boxes to maintain their efficiency in organization, another human characteristic.

Boxes facilitate thinking as well. They are a form of physical organization allowing my wife to keep track of a variety of family pictures dating as far back as, . . . oh well, she says, "never mind". I use boxes to keep track of my receipts for tax purposes. I am sure that my receipts from seven years ago are in the orange and black shoe box. However, I am not clear, at the moment, where that particular box is. In the garage, I'd guess.

Boxes serve two main functions in science. First, boxes contain things. Things must be contained or they spill out everywhere. Since science is the study of things, it always has materials spilling everywhere. I recall that my late friend, Dr. Walt Kelley, maintained his grass collection in boxes at the University. Dr. Kelley was highly organized, and his boxes were never disorganized in any way. They were never even allowed to be out of alignment on the shelves.

I know this for a fact as I used to see how little I could move one box before he would notice and make realignment. As far as I could tell, it was about one sixteenth of an inch. I felt bad when I later discovered that little things like that upset him, just not bad enough to stop. I curtailed the cruel practice as soon as he learned it was me and retaliated.

It might surprise you to know that even empty boxes are of use to scientists, at least metaphorically. While science is the study of material things that need to be contained, the end result of most scientific studies is to generate abstract laws or principles which are not material at all. They are ideas.

Hence, the study of falling apples led to the law of gravity, a non-material idea. The study of water displacement by objects led to the idea of density. One cannot hold gravity or density in a box. The box would appear empty though it would contain all the knowledge gained from falling apples and displaced bath water.

However, empty boxes that are crammed with old ideas are like saving empty boxes for when you need them. When we need to know about density, we can get the idea out and use it. Metaphorical boxes are almost as useful as doggie bags, which are almost always boxes.

Just consider how humans use boxes. There are jewelry boxes, cardboard boxes, tool boxes, electrical boxes, fuse boxes, post office boxes, penalty boxes, Jack-in-the-boxes, lunch boxes, shoe boxes, Pandora's box, signal boxes, first aid boxes, and lock boxes. In other words, humans need boxes in which to store stuff.

All this doesn't even address getting boxed in, getting one's ears boxed, or going to a boxing match which takes place in a boxing ring. Hey, maybe all this explains why there's no room in the garage for the truck!

It doesn't explain why we think we need all the stuff.

## BALTHAZARS FEAST

I dreamed I saw the finger write at Balthazar's feast.
I watched in foreboding until the writing ceased,
As if I were he, or he were me, and the message both haunting.
Our kingdoms are numbered, and both are found wanting.

I dreamed I saw the finger write at Balthazar's feast.
On the wall like the morning, sky silver in the east.
Like the King who knew things were not as they seemed,
I know not the interpretation of my dream.

I dreamed I saw the finger write at Balthazar's feast
While being served in the vessels of the priests.
The finger wrote on in holy silence,
"Thou art found wanting when placed in the balance."

I dreamed I saw the finger write at Balthazar's feast,
"The last shall be first, and most shall be least."
To your kingdom, and his, and all that you've added,
Tonight it is lost and the remainder divided.

# 7

# BEGINNING UNDERSTANDING

*"Alas! how easily things go wrong!*
*A sigh too deep or a kiss too long,*
*And then comes a mist and a weeping rain,*
*And life is never the same again."*
*George Macdonald*

I have no idea how something begins. If you think about it, which I try not to do because I don't know how to get started, there is always something that happened before whatever is beginning that started the thing that is beginning.

For example, do I awake each morning because the alarm goes off or because I set the alarm the night before? My wife often asks why I set the stupid alarm anyway. She hates alarms.

So what begins a beginning? This question is especially important in science. Scientists puzzle about the beginning of it all, although most scientists can't tell you what the actual beginning was. Why does a person make the seemingly-illogical decision to become a scientist?

To begin to become a scientist, with the union card that says "PhD", a person has to subject themselves to years and years of mastering arcane, complex, often-non-intuitive, almost-always-mathematical, obscure, concepts about a smaller and smaller piece of the world. I am told there are psychological theories about why someone would do this, but I don't know how they got started either.

This conundrum has practical implications for even those who aren't scientists, because of all the things in our fascinating world to see, do, and be interested in, why do people begin to do the things they begin to do? I mean a person could just as easily get interested in moving medieval-looking pieces around on a checkerboard as in hitting a little white ball with a club. Today, people are suddenly interested in cooking shows. Yet, my mother tried to get out of cooking whenever she could.

Humans give beginnings a set time. They're like birthdays. We celebrate a birthday and ignore all the interesting things that led up to it for our parents. We say that we declared our independence from England on July 4, 1776, but of course, we had become independent in our minds long before that. July 4th is just the date when we finally worked ourselves up into doing something about it. Maybe todays politicians should consider that.

Just to confuse things a little more, the end of something is always the beginning of something else. I've heard of the "beginning of the end" before. Have you ever heard of the "end of the beginning"? Somehow that seems redundant.

**BEGINNING UNDERSTANDING**
Born seven years before the wind,
In the valley of the sand,
Beneath southwestern sun.
Yet I do not understand
How darkness folds on darkness
With the coming of the night
And the full moon brings cold winter
With borrowed, reflected light.

I've heard it said by dreamers
That, clearly it's a sign,
When lightening casts a shadow
And night birds call in rhyme.
A sign of things to come,
Of things yet unborn,
And things that only come
In the coming of a storm.

Like the life that comes touched by stars,
Or the life that ends forlorn.
All the sons and daughters
That God causes to be born.
And this is such a night,
The kind one often sees
When great men are born and die,
While plain men are on their knees.

Life is to fulfill life on earth
In the meridian of time,
When shadows come to life
Precept on precept, line upon line.
In a room filled with shadows,

Spending their last night,
The baby and the old one
Each beginning a new life.

# 8

# BEYOND THE VEIL

*"Her heart - like every heart, if only its fallen*
*sides were cleared away - was an inexhaustible*
*fountain of love: she loved everything she saw."*
*George MacDonald*

Memories have to have a certain structure to them. There is nothing to see in the brain except neurons and synapses which act mostly like electric wires, switches, and circuits. So memories must be made by these things. Scientists have looked pretty closely and haven't found any filing cabinets, shoe boxes, or even safe-deposit boxes.

Because we are conscious of memories, it appears that they are all stored in the top of the brain, the cerebrum. Everything that happens below that level is pretty much unconscious. My wife isn't sure on this point. She thinks my unconscious level goes all the way to the top. I'm actually pretty proud of that since it's about the only thing I have that's ever been tops.

The cerebrum is approximately six cell-layers thick. That is about the thickness of a stack of six playing cards. However, this part of the brain is estimated to hold about 20 billion neurons, each with switches to about ten thousand other neurons. Well, that's only if you are an adult. Very young children have fewer switches but actually about the same number of wires.

So if we remember something, what is it that's happening in the cerebrum? Some wire must be activated and a message sent to another wire through a switch. This second wire continues to transmit the message to the next switch, and so on, until it reaches some point that has been reached before.

However, it's even more complicated than that. If there are twenty billion neurons connected by ten million switches, it means that any given end point could be reached by an almost infinite number of pathways through this three-dimensional fish net or matrix.

If the end point, which may be a specific event or fact, can be reached by many different pathways, some of those paths are likely shorter and faster than others. Thus one might recall some things, especially those recently learned, slowly. The pathway to the end point being especially circuitous. If the memory involves a physical response, we might be even slower to respond because different pathways in and out may take different and indirect routes.

However, we know that people who continue to recall specific things generally become faster in response. The only way I can see this happening would be if after repeated responses, certain switches became enlarged or more efficient in some way. This would create a "favored pathway" which would grow stronger as it is exercised, thereby increasing efficiency and speed of recall.

Something like this must eventually happen when you become adept at a specific activity like playing an instrument. The reinforced pathways are accessed so rapidly that the hands can respond almost instantaneously without actual, conscious thought or needing to identify the end point by name.

What if we created a pathway, very early in life, that identified the ideas of distance being divided into smaller segments, something like a number line, or a thermometer, or a musical scale? Would it later make learning easier to use that wiring and those switches to recognize other things separated by degree like radio frequency, light spectrum, or three dimensional arrays?

And what if you had a memory line for a musical scale and later had to work with the idea of light intensity? Do you suppose you might hook the idea of lumens onto the musical scale line since they are similar? It seems like learning in this way would be an efficient way of doing things.

It seems reasonable that children, with an organized set of experiences in early life, might be better prepared for creative and logical thought later in their academic work. Children with more haphazard and multifaceted childhood experiences might later struggle with organized or creative thinking simply because of inefficient "favored pathways".

I wonder if all our experiences, in mortality, might not contribute to building memories or pathways, that will be of use after we die in still higher callings and advanced learning. Perhaps we will recall, not just events, but we will use those efficiently developed pathways while on earth to better execute our tasks in the hereafter. Anyway, I hope I don't forget the following story.

**BEYOND THE VEIL**
There was a dream, or I thought it so.
But sometimes it is hard to know.

After night, followed by day,
Dreams sometimes slip away.
At least it seems so now.
That sweet kiss upon thy brow;
Was it real or does it seem
Like it might have been a dream?

Still, let me now tell the tale
Of things that happened beyond the veil.

I remember standing on a distant hill,
Wet with sweat and windswept chill.
I waited long for you to come.
Waited 'til my heart grew numb,
And I grew restless, so I went to see
What was keeping thee from me.
The ocean roared miles below.
The wild wind whirled to and fro.

Along the cliff and rocky trail
I watched for you beyond the veil.

We parted just the night before.
Your brow I kissed at your door.
And swore that we should meet again
The following night, and defy all men,
On the cliffs above the sea.
And from all this we would flee,
From those who said we could not be,
Sealed together for eternity.

In the face of fate mans will is frail
On the other side of the veil.

The horizon that had been lost in mist
Now pounded on me like a fist,
And water ran everywhere
Across the rocks in lightnings flare.
As I made my way along the edge,
Through sodden grass and dripping sedge,

Towards the trail that led to the sea,
And the ship I had prepared for thee.

Then far above along the trail
I saw you standing there through the veil.

On the cliff against cloud-swept sky,
Your silhouette filled my eye.
I marveled when I saw you there
That such, for me, you would dare.
With haste I clamored to thy side
And embraced my coming bride.
We began the treacherous descent
To the sea of our ascent.

Across the sea through stormy gale
We two escaped beyond the veil.

There was a dream, or I thought it so,
For sometimes it is hard to know
After night and followed by day,
Dreams sometimes tend to slip away.
But now it's clear it was no dream.
Storms and trials were as they seemed.
All of it was meant to be.
You and I for eternity.

So there you have it, the very tale,
Of things that happened beyond the veil.

# 9

# BYZANTIUM

*"Whose work is it but your own to open your eyes?*
*But indeed the business of the universe is to make*
*such a fool out of you that you will know yourself*
*for one, and begin to be wise."*
George MacDonald

"They seeing see not; and hearing they hear not." Anyone who has had children understands this Biblical saying perfectly. Every parent has told to a child to take out the trash and the child claims to not have heard them. My wife tells me that I am extremely spiritual in this way. Actually, this saying has always been a mystery to me. However, scientifically, it does make some sense.

Have you ever noticed that our eyes are on the front of our faces, our ears on the sides, and our nose and mouth are sort of smack dab in the middle? It's not this way for all animals, you know, although I can't think of any animals that have noses on the side of their heads. Well, maybe sharks, but that's another column. However, some animals do have their eyes on the sides of their heads and their ears facing the front. Horses do, for example.

I have sort of taken seeing and hearing for granted. I can still hear just as well as I ever could. It's just that now I can't understand the words. Our modern education system doesn't teach the youth to speak clearly anymore. Even the people on television mumble more than they used to. With each year, my understanding of those who hearing, hear not, increases.

The anatomy of hearing and seeing depends on how animals use their senses. Humans tend to think the senses are for pleasure; seeing beauty, hearing music, smelling perfume, the feel of clean sheets, or tasting a steak. However, our senses are far more important these things suggest. Our senses keep us alive!

Senses are used for sampling the world around us and informing us of changes that are taking place. While some such changes are pleasurable, it is far more important to note when the changes are detrimental. For example, flashing lights and sirens may be alerting you to detrimental changes about to take place in your life.

The science of making sense of the space around us is based on triangulation. Triangulation is the process of determining the distance to an object by measuring the length of two sides of a triangle and deducing the angles and length of the other side by observation from a baseline. Yeah, I know, I hated trigonometry too. However, we use it every day whether we know it or not. That's actually my favorite kind of math; the kind where I don't know I'm doing it. It's so much easier than thinking.

The two observation points of a triangle form the base line. If you want to pinpoint the position of an object accurately, you need to see it from two different positions so you can triangulate it. For some animals, it is important to have accurate depth perception. In those cases, it is better to have sensory organs facing forward. In this way, they receive information from separate positions a few inches apart.

However, what if it is more important to have a wide field of vision in order to perceive a predator sneaking up from behind or the side? Then it is better to have sensory organs on the sides of the head. The positioning of the eyes on the sides of the head provides the animal with nearly full circle vision. The tradeoff is that they literally can't see their hands, paws, hooves, whatever, in front of their faces!

Obviously a lot of animals combine sensory systems. Horses, having eyes on the sides of their heads, have difficulty telling how close they are to an object. On the other hand, their ears generally point forward. Humans have eyes in front and ears on the side. Interestingly, even though our nostrils are not very far apart, we are still pretty good at locating unusual odors.

Of course there is another entire way of thinking about seeing and hearing, and that is in the use as an allegory. What do you say to people who cannot seem to see the advantages they have right in front of them, or those who make foolish decisions without being able to appreciate the consequences?

Such was the case of the people living in the ancient city of Chalcedon, set on the Asiatic side of the Bosphorus. The Bosphorus is the opening from the Sea of Mamora into the Black Sea. Chalcedon was settled in a low area where a river flowed into the narrows. The European side of the Bosphorus was at a higher elevation and controlled the narrow passage way. So when a Greek Byzos settled there on the advice of his soothsayer, the soothsayer supposedly told Byzos that the people of Chalcedonia must be blind not to recognize the superiority of the European position for controlling the water way.

Thus, Byzantium eventually replaced Chalcedonia and was known as the city of the blind. Later Byzantium became Constantinople, and still later, Istanbul. But the people of byzantium are an example of a people who could not see their own advantages. (This poem is questionably dedicated to the people of the United States of America here in the early twenty-first century. They could not see the good in the country they had created and sought hope in a system that had demonstrably failed.)

## BYZANTIUM

City of the blind resting on a hill,
Those within its walls could not see
The advantage of their place
And do not see it still.
      Only time will reveal
      The conqueror is real.

City of the blind, at the golden horn,
At the mighty water dividing land from land
Where Constantine's influence reigned,
Now of his power is shorn.
      Only time will reveal
      The conqueror is real.

City of the blind, too lovely to sustain
Music from those golden domes,
As if from Heaven's gate,
And spread across the world like rain.
      Only time will reveal
      The conqueror is real.

City of the blind knew no bounds.
Your dream sustained by light
When mighty rulers attacked at dawn,
Bayed like braying hounds.
      Only time will reveal
      The conqueror is real.

City of the blind, a light from the sky
Repelled invading hordes.
You hailed salvation as the light

While the end of time drew nigh.
  Only time will reveal
  The conqueror is real.


City of the blind now reveals
The frailty of consequence.
Once again man has failed to see
That conqueror time is real.
  Only time will reveal
  The conqueror is real.

City of the blind, with faith in man alone,
You forget the source of light
That one time saved you from the dark.
That leaves you overthrown.
  Only time can reveal
  That Emperor time is real.

# 10

# CEMETARY PASSAGES

*The best preparation for the future is the present*
*well seen to, and the last duty done.*
George MacDonald

I wonder if we'll know when we are dead. We don't get to know when we will die, so I wonder if we will know it has happened afterward. I like the calendar. It's comforting to know when the end of a year will come. It's always after 365 days. You can prepare for something like that. It's predictable.

I could tell that my students were able to anticipate the ending of my lectures. I wasn't as precise as the clock or the calendar, but I hardly ever went over time because the students started packing up long before I was through talking. It was a relief to realize I was about finished.

I guess everything comes to an end. As far as I can tell, it isn't the endings that bother humans. It's the difficulty of predicting the endings. Well, that and the fact that many endings get kind of long and drawn out. That's why the calendar is kinda cool. You know when the end is coming, it's over in an instant, and you know it happened.

Someday we all die. Of course, on all the other days, we all live. It's not a bad trade-off really. So far, I have lived over 25,000 days. I'm being purposefully vague here, but I do have only one day of dying still to come.

I don't like the idea of bucket lists. They would make me feel all rushed to get things done. I would be really ticked off if I got everything done and then found out I had to sit around for years waiting to die. My personal goal is to take things nice and easy up to the day I die and then kind of taper off. It would be convenient if we could know when we need to have our affairs in order. Of course, some of us would just put it off to the last minute anyway.

What's strange is that all living things are made of the same basic stuff, but they have different life spans. How come the tortoise gets 73,000 days, and humans get, on average, a third of that? Maybe I need to slow down.

It turns out that the difference is in the number of heartbeats we are allotted. Way back in the 1930's, a Swiss named Max Kleiber studied the relationship between mass and metabolism. He came up with a formula that predicted the energy burned per unit of weight is proportional to an animal's mass raised to the three-quarters power. I hate fractional exponents! It's a lot simpler to say the smaller you are the more calories it takes per ounce to keep you alive. This should be encouraging for the tinier humans during the holiday season.

However, if you plan to eat a lot of food fast, you have to metabolize it quickly so that you will have room for the "more food" you are going to eat. Metabolizing food requires energy and generates heat. Getting rid of heat and circulating energy requires heartbeats.

It turns out that, at least for mammals, there is an allotted number of heartbeats per lifetime: about a billion. A relaxed shrew, which may be an oxymoron, has a resting heartrate of about 850 beats per minute. At that rate, they have a life expectancy of about two years. Some whales have heartrates of ten to fifteen beats per minute. This buys them about two hundred years.

I figured that, on the average, my heart rate has been about seventy beats per minute for most of my life. Of course there are those occasions when I did stupid, terrifying things that increased the rate, and there is my wife who still takes my breath away. I figure those things are offset by all the hours at work during which I didn't really do much of anything. Seventy is probably a good average. So if you calculate this out, at seventy beats per minute for a billion minutes, I'm already dead. I have been since I was about twenty-seven years old.

## CEMETARY PASSAGE

Snow lay like a cloud across the rocky field,
With stones protruding like grey sky,
Making the world seem upside down.
While the ancient bell tower peeled,
The sky looked more like clay
Than like the earths shining crown.

The epitaphs are weathered and blown away.
Who are the ghosts of this town
That scurry through the misty streets?
I'm afraid I have arrived too late to cry
Though I came as soon as I could.
Nor do I know under which stone they lie.

News travels slowly when the news is sad,

And it was late in the day when they told me.
When I complained, they hastened to say
That news travels slowly even if it's glad.
It was so far and had been so long,
Over distance and time, I had to find my way.

At dawn I arrive at the old church door
That stands open on a crooked hinge.
The walls are gashed and scarred,
And the rooms have broken glass on the floor.
The one long hall is musty with dust
With just the frame of swings left in the yard.

I made a casual nod to a passing visitor
While making my way, searching the rows.
No mumbled greetings there in the snow
And his passing but a blur.
Brushing snow from each stone I pass,
Up and down each row I go.

Snow lay like a cloud across the rocky field.
I kick and scuff the snow away
To find epitaphs buried with the dead,
Until I find the one sought and kneel.
I Brush snow away with cold fingers.
Ben and Sadie is all it says.

It's as if the storm had blown their dreams away.
No sign of life, or dream, or place.
Bodies buried by dirt and cold.
Their time buried like the clay.
The part they played on the earth
Is all their story left to be told.

I feel the ghosts around me now
But cannot make out what they say.
I think they say it's time for me to go.
With furrows of wonder on my brow,
I look from whence I came,
And see no footprints in the snow.

# 11

# COME WALK A LONELY TRAIL WITH ME

*"It was foolish indeed - thus to run farther and*
*farther from all who could help her, as if she had*
*been seeking a fit spot for the goblin creature*
*to eat her at his leisure; but that is the way*
*fear serves us: it always sides with the thing*
*we are afraid of."*
*George MacDonald*

Humans are all confused about health. They don't actually know what it is. Usually we say something like "Health is the absence of disease." But we don't know what that is either, other than the absence of health. Still we manage to spend several billion dollars on disease prevention and health promotion every year.

The words health and disease are really impossible to define because they are disguised value labels, not actual objects. Health is something we think is good, and disease is something we think is bad. If I am hurting in some way, I think it's bad. If I feel rested and energetic, I might say I feel good. If I touch something silky and clean, I might say it feels good. Touching something sharp might feel bad, or not, depending on how and why it is touched.

Good and bad may have different interpretations, and something might be both good and bad all at once. Surgery may be unpleasant. But during surgery a cancerous growth might be removed, which is good. Having sickle cell anemia may protect you from getting Malaria if you happen to live in a malarias region.

The point is that the terms health and disease, and good and bad, are value judgements. They are not scientifically-defined, material objects.

The choices we make in life are, much of the time, simply value judgements. The question, "Should I go to college, or not?" may actually involve a decision about what you think you might want to be doing in a future that may not even be there. The future will almost never be like it is now, so your decision is really about what you value more than what will actually occur.

Basically, humans have only two decisions to make. Do we value wisdom and experience, or do we value safety? We can choose to venture forth into the unknown future and embrace the learning, the experiences, and the heartaches.  Or we can hunker down and live long, uneventful, safe lifves with our progress and learning dammed up behind our ignorance.

This is the choice Adam and Eve faced.  Is it better that humans suffer so that men might be?  Or should humans partake of the tree of life and live forever in their sins, progress forfeited?

## COME WALK A LONELY TRAIL WITH ME

Come, walk a lonely trail with me.
Walk quietly and be free.
The crooked path, with no human sound,
But fairy rings all around.

The trees stand guard against the way
And force all within the path to stay.
Limbs and leaves block the sun
And the moon when nighttime comes.

There is a man within a tower
Alone for years and days and hours.
He waits and pines the years away
As he longs for his appointed day.

A door swings open in the breeze.
The front path strewn with rotting leaves.
What once was crafted magnificence
Is now a fading transience.

For no soul went within, without.
The bricks had mortar falling out.
Vines climbed forgotten walls.
Moss grew like ancient scrawls.

Twin towers on each side stood
To guard the arch that led to the wood.
One must take care when going there.
Of all things magic, be aware.

Within the wood a gnarled tree,
With limbs twisted like broken knees.
The tree of life with barren fruit,
Broken limbs and twisted roots.

No other tree grows close by.
Those that tried have all died.
The grass is brown and dry as dust.
The soil is cracked with broken crust.

Yet the foliage is green upon the lawn
Around the castle in the dawn.
Fed by dew and morning rains
And marred by spots of dying stains.

The man knows well his eternal fate.
He stares without through an iron grate.
For he defied the flaming sword.
The words of warning he ignored.

Of the tree of life, he partook.
Ignored the warning in the book.
From his beginning there is no end.
He must live forever in his sins.

Come, walk a lonely trail with me.
Walk quietly but be free!
The crooked path, no human sound,
Yet fairy rings all around.

# 12

# FALLEN ANGEL

*"And her life will perhaps be the richer, for*
*holding now within it the memory of what*
*came, but could not stay."*
*George MacDonald*

Some people don't believe that we live again after death. I think we do. I can't imagine there being nothing. It's like trying to imagine the infinity of space. If space is empty, what is just beyond the empty?

Mark Twain once said, "When I was young I could remember anything, whether it happened or not." I wonder if, when we are dead, we will be able to remember what life was like when we were alive. Obviously, if you don't think we'll exist after death, you don't believe you will be able to recall what went on here. But if you do believe in life after death, it would seem that we would have to remember what happened here or what would have been the point of ever being here?

Some people don't believe we lived spiritually before we were born. But some people do. I think we did; I just can't remember it. William Wordsworth thought so too, but he was just another poet. In his poem, "Intimations of Immortality", stanza 5 he says:

"Our birth is but a sleep and a forgetting:
The soul that rises with us, our life's star,
Hath had elsewhere its setting,
And cometh from afar.
Not in entire forgetfulness,
And not in utter nakedness,
But trailing clouds of glory, do we come
From God, who is our home:
Heaven lies about us in our infancy.

Like he says, though, our birth is but a sleep and *a forgetting.*

What if we could remember what went on before we can remember? Would things make more sense here on this earth? I think so. However, it would also take away much of the wonder and awe which we experience in mortality.

Most people think the brain tells the hand what to do. However, it seems to me that the brain doesn't know much about the hand until the hand tells it something. How can a brain know what is hot or cold until the hand has touched hot and cold things? How can a brain know left and right until the hand has learned to put the right shoe on the right foot? Okay, technically I suppose, that is your foot telling your brain what is right and left. Still, the concept is the same thing.

I bet I know what the first thing was you did when you woke up this morning. You hit the snooze button with your hand. Early morning is probably the only time of the day when I have a lot of time on my hands. You may be one of the rare, disciplined, disgusting, individuals who actually gets up when you are supposed to. Then the first thing you did was turn the alarm clock off, so it wouldn't disturb the other more normal people in the house.

The second thing you did was use your hands to push back the covers so you could get up. In fact, you probably used your hands for dozens of things long before you uttered your first word. Okay, there might be a few maladjusted folks who say things when the alarm goes off, even before hitting snooze. But that is a learned response and must be carefully taught! Speaking before thinking is actually a dangerous thing to do and should only be attempted by specially trained windbags called college professors. Do not try this at home without supervision.

Babies reach for things long before they learn to talk. Most are even successful at picking things up and putting them in their mouths before uttering their first words. Could that be the origin of the phrase a "hand to mouth existence"? I am told that about a quarter of the motor cortex of the brain, the part that senses and controls the body, is devoted to the hand. You gotta hand it to the hand if it is that important!

I am taking that last fact on faith since I didn't have a hand in figuring it out. I don't know how someone could know about the motor cortex unless they actually had a hand in determining it. I'm not sure we can really know anything unless we put our hands to it. On the other hand, you have different fingers.

What does all this have to do with science? The history of science and human invention, which each generation likes to call progress, occurs mostly through understanding that we develop the physical world. We mostly learn to understand the physical world through our hands. Like Jacob Bronowski observed, "The hand is the cutting edge of the mind."

While science has amassed a large catalog of facts, it is not the collection of facts that comprise science. It is true that we can use the stockpile of knowledge to help us do things. Science, however, is the making of new facts through direct experience with the physical world. This often requires doing, or making new things, to better manipulate the world. Or we are required to use previously developed tools to understand the world.

It is hard to understand computer code if one does not understand electrical switches. It is hard to understand hearts and lungs if one does not have some experience with pumps, pressure, and liquid flow. So what do we Americans make nowadays? Does it make any difference what we make? Do children make anything in school after grade four? Can a college graduate actually make anything? Maybe in today's world it would be good if we could find ways for the hand to teach our brains something new. Or we could say, "Allow your hands to teach your brains something new.

If we lived before we were born, I presume we didn't have any hands. So even if we could remember what it was like then we still wouldn't know anything we need to know to survive here, now, like how to use our hands. Maybe that's one of the purposes of life, to learn from our bodies.

Anyway, I think I remember this tale, although it is sort of like a dream.


## FALLEN ANGELS
One evening high on the mountain,
On the bank of a cold gurgling stream,
I lay on a rock warmed by the sun
And watched for the early moons gleam.
When I saw angels falling from Heaven,
Floating to earth far below,
I wanted to run and to hide
But wasn't sure where I should go.

One angel came and stood before me.
Her equal never seen before.
Dressed in exquisite attire,
Snow white was the mantle she wore.
Her eyes were like sparkling diamonds
Or stars on a cold frosty night.
Her hair hung with hemlock and roses.
She glowed with a heavenly light.

I cried out, "Save me, save me forever!
Stay! I can't let you go.
Without you I will grow weary.
Without you I never will know."
She smiled and her teeth were like pearls
"Come find me," was all she replied.
Her voice like soft wind in the forest.
"Come find me," was all that she sighed.

Then I dreamed that I awakened.
Or perhaps I awoke from a dream.
The rock had grown cold in the night,
And the moon was all to be seen.
The angels no more were falling.
Things were just as they were before.
The pretty, fair angel had left me
Alone on the mountain stream shore.

Each year I returned to the mountain;
To the rock, and the stream, to it all.
To wait for the moon to arise,
And watch for the angels to fall.
But, year after year, it is wasted.
I wish I could slumber again.
To see her with sunbeams around her,
I would never more awaken then.

At last I grew restless and weary.
I thought, "My vigil is in vain."
Awake, or a dream, is a riddle.
To think that I know is insane.
So I returned to my home in the valley,
Spirit, or alive, I don't know.
I only go to wherever they send me.
There is no other place to go.

In my new home, I drifted and wandered.
It seemed it had been a lifetime
Since a time vaguely remembered
When I saw an angel so fine.

Then one evening, high on the mountain,
On the bank of a cold gurgling stream,
I lay on a rock warmed by the sun
And watched for the early moons gleam.

When I saw an angel fall from Heaven,
Floating to my valley below.
I wanted to run, to embrace her.
But didn't know if I should go.
She stood there, right before me,
Her equal never seen before,
Except for the angel on the mountain
In the snow white mantle that she wore.

Yet different from there on the mountain,
Less angelic, yet maybe more.
Different than the angel I'd seen,
A woman stood on the shore.
I cried out, "Save me, save me forever!"
Stay! I can't let you go!
Without you I will grow weary.
Without you I never will know."

She smiled and her teeth were like pearls
"You have found me," was all she replied.
Her voice was like soft wind in the forest.
"You have found me," was all that she sighed.

# 13

# GUIDED BY THE CONSTELLATIONS

*"It matters little where a man may be at this
moment; the point is whether he is growing."*
*George MacDonald*

Since my retirement from the University, I no longer have large research grants and extensive laboratory space for my research. Well, actually, I've never had any large research grants, and most of my research had to be done in the basement of the old Wubben Hall because all lab space was needed for classwork. Oh, yeah, there is a basement in the old Wubben Hall. You don't want to go there.

Anyway, I now have time to speculate on some less-important scientific questions that have defied analysis. Like, "What are people for?" I am always being asked, "What are mosquitoes for?" or, "What are centipedes for?" I think what people mean is, "What good are mosquitoes and centipedes to people?"

I have read that some think there are too many people. How can we decide there are too many people unless we know what people are for? If the job people are for is getting done, and there are people with nothing to do, then I suppose we should get rid of a few of them. I have a couple of suggestions.

I'm reminded of the time, several years ago, when the government was telling us that there were too many farmers. We had to get rid of some of them, so we could be more efficient. Interestingly, I'd never heard a farmer express that concern. Too, I'd never heard a professor of agriculture opine that there were too many professors of agriculture. However, if we had fewer farmers, wouldn't we need fewer professors of agriculture, and a smaller Department of Agriculture?

Where did all those farmers go anyway? They went to cities, of course. Apparently our cities need a lot more people than our farms to do whatever "people are for". I sort of know what people are for on a farm. I am not as sure why we need people in the cities, but there must be some reason because all the farmers went there.

Today it is starting to look like we have too many people in the cities. With all the efficient farms, automated factories, labor-saving devices, robots, shorter workdays, longer vacations and early retirement, it appears that people are just for goofing off. My wife thinks that's what I've been doing for years. . .

Maybe there are both too many people in the cities and on the farms. If we don't know what people are for, it's hard to tell. Of course, if we think there are too many, I suspect there are further inhumanities on the horizon.

Then again, I may be going about this all wrong. Maybe I should be concentrating on the question, "What good are people to other people?" If we judge other living creatures like mosquitoes and centipedes by this standard, it seems only fair to judge people in the same way. Several testable hypotheses come to mind.

If it weren't for the people who invented computers, I would still be writing my columns in long hand. Therefore, my editors would be unable to read my columns, so you would never have to suffer through this stuff! Okay, I guess that could be taken as a pro or a con.

But without people, we wouldn't have electricity, cars, air planes, indoor plumbing, television, cell phones, or Facebook. Nor would we have Shakespeare's plays, Rembrandt's paintings, or Beethoven's symphonies. Ideas and creations come from people. In fact, as far as I know, ideas never come from anything but people.

There is at least one other thing that humans "might be for". Almost all people are fellow humans, friends, neighbors, parents, children, citizens, children of God, and kind people. So my scientific hypothesis is that people are for taking care of each other. When people use their abilities and creativity to care for each other, then it stands to reason that we need more people.

## GUIDED BY THE CONSTELLATIONS
A gentle throbbing inside my skull
Has a strange magnetic pull.
The feeling is vague, but the direction clear
Like the guiding stars of the hemisphere.
Forgotten awake, but in my dreams,
Like a slow moving and hidden stream.
So I stuff those dreams in a canvas sack
And promise myself that I'll be back.

I'm guided by the constellations,
The star that guides my Father's son.
I wonder if she sees the same.
I wonder if I'll know her name.
I wonder if she'd ever be
Waiting for a boy like me.

The sky above the merging rivers
The color of unpolished silver.
The town below is a stain of mourning
The faded color of ancient mornings.
Now that I wake from my dream
Neither appears like it seems.
So I stuff the scene in my canvas sack
As soon as I find her, I'll be back.

I'm carried by my muscled thighs.
Guided on by sightless eyes.
I wonder how she finds her way,
Or if she decided not to stay.
I wonder if I'll be the man
Who can win her heart and hand.

With legs of lead and lungs of fire,
In the distance I hear a choir.
Singing eye of newt and brain of toad,
A fearful odd melodious ode.
Yet other strains assault the ear
Of more Godly hemispheres.
So I stuff my guitar in a canvas sack
And struggle on to make it back.

I depend upon wounded feet
To save me from these mortal streets.
I wonder how she found her way.
I wonder why she chose to stay.
Still together at the end,
Bound together eternal friends.

Clouds occlude like cataracts.
Light is rapidly fading black.
Holding hands, I'm not afraid.
This, the life, for which I prayed.
Hard to see but it's becoming clear
That up ahead is a funeral bier.
So I stuff my memories in a canvas sack
And heave a sigh, I made it back.

-14-
# IN MY DREAMS

*"We must do the thing we must*
*Before the thing we may;*
*We are unfit for any trust*
*Till we can and do obey."*
George MacDonald

The problem with complex stuff is it's complexity. That's the part I always hated about science. If something is simple enough, I get it. If it is too complicated, I tend to get a peanut butter and honey sandwich instead. I understand that.

Most of what happens in complex systems can't be visualized because the processes are too complex. However, that means no one else knows how it works either. So I can imagine a thing working as simply as I want, and no one can really say otherwise. Of course, when it comes to how I imagine systems working, no one really cares anyway.

For example, no one really knows how the economy works. Anyone can just make up an explanation, and his idea is as good as anyone else's. That's probably not the best example, though, because economics isn't a real science.

Maybe I should use ecology as an example. Ecology used to be a science before it became a political movement. It's pretty complex, and no one really knows how it works either. That's why politicians love it.

The difference between complex things and simple things, as far as I can tell, is quite complicated. I am told that complexity is a system of multiple components that interact with one another in such a way as to dissipate energy. When such a system exists and you turn it on, weird and unpredictable things happen. That sounds a little like a good marriage. That, or the interstate at rush hour.

The point is simply that I have lost track of the point I was trying to make. But I shall go ahead and make a point. I like to deal with complex issues because no one else understands them either. Therefore, I can't be proven wrong. An added bonus is that I don't have to be responsible for any outcomes.

In our family, I take care of the complex things because I am a scientist. I tell my wife what we should do about ISIS, the national debt, and the educational system. She handles the simpler things like where we live, our investments, and what I get to do all day. Her list is longer, but my responsibilities are more complex. So it just about evens out.

Here is how I imagine the difference between complex things and simple things. If you can pretty accurately predict what will happen under certain conditions, it's simple. If you can't predict accurately, it's complex. Like if you introduce your sister to your best friend, what might happen? There is no telling. You might know a great deal about each of these people. But when you actually bring them together, you have no idea what that mix might bring. That's complex! That also explains why there are so many failed blind dates.

If humans were to introduce a new species into an environment, what would happen? It's unpredictable. If we actually reduced carbon dioxide emissions worldwide, what would happen? Nobody knows. Maybe we'd all freeze to death. What if we actually ran out of Peanut Butter? That's why I took up beekeeping - to eliminate the possibility of running out of honey. I delegate the peanut butter supply to my wife's keeping though. I can't take care of everything.

Scientists actually know a lot about complexity. There are several books written, and whole research institutes devoted, to the study of complexity. I am thinking of starting a Foundation to study simplicity. I hear foundations can be lucrative, and I don't think anyone else is actually trying to be simple. My wife says I am a natural.

What is interesting about this discussion is that, notwithstanding the tremendous advantages and improved living conditions science has given us, the things we can control and predict are still the relatively simple aspects of the world like electrons, gravity, DNA, continental drift and such. No one really understands why their spouses married them in the first place.

**IN MY DREAMS**
In my dreams, I've seen my Father,
The man I will never be.
All around are the shapes of men
From whom I struggle to be free.
There's the ghost of Mother's dreams,
The man I'll never be.
In the shadow of the wind
Is the only place he's seen.

In my dreams, I'm unfamiliar,
Like a man without a past.
Twisted memories in a mirror,
Like dark shadows on a glass.
When I laugh, I dream my Father.
My Mother's dreams are shifting sand.
As I reach to draw them in,
They run like water through my hands.

In my dreams, I see the future.
It is so much like the past.
The future seems just like we were
Approaching and receding fast.
Carried forward like the wind
To a place that's never been.
I'm the man in the mirror
But the details are just a blur.

In my dreams, I see the past.
They say things never change.
Seventy years before the mast,
Travels beautiful and strange,
Forever locked from me now.
Useless land of used to be,
Years enslaved to the plow,
Now I'm yearning to be free.

In my dreams, I see another
Of the faces I have worn.
There are sacred signs and wonders
From the time that I was born.
That time is gone, but I am haunted,
Only the present clearly seen.
Past to future I am seeking,
A restless go between.

# 15

# NO PLACE LEFT TO GO

*"Only he knew that to be left alone is*
*not always to be forsaken."*
George MacDonald

I sometimes wonder about how our thoughts originate. Would it have made any difference if we'd thought otherwise?

For example, why do we think that light must come from a source? We all believe that the sun emits something called light. So why didn't we think of the sun as being a big, dark-sucker that sucked up all the dark. Wouldn't that have made just as much sense? There could even be many dark suckers out there, each with varying abilities to suck up the dark. If the world were filled with dark, it would take a really powerful sucker, like the sun, to suck it all out. The moon and stars would be less powerful suckers that couldn't quite get it all.

Perhaps light bulbs, too, have various capacities for sucking up dark. Within their immediate vicinity, they can pretty much get them all. But a few feet away, they just don't possess enough suction. Then when we turn off the sucker, the dark fills up the space again.

It could work something like this. Maybe there are dark particles called darkons that have something like an electrical charge and are attracted to the opposite charge. What we call light sources would actually be oppositely charged "lightons", and they would attract darkons. Wouldn't everything still work about the same way if that were thought to be the case?

Here is another one. Why did Newton think there must be an attraction between two bodies that are roughly related to size? He based his thinking on the principle of acceleration. Since velocity of a falling apple changes from zero as it is hanging on the tree to something else as it falls, it must have a force accelerating the fall. He imagined that if the apple tree were twice as high, he could expect the apple to be accelerated even more by this force. He then began to realize that this force extended far beyond an apple tree.

So he came up with the law that says, "Every object in the Universe attracts every other object with a force directed along the line of centers of the two objects that is proportional to the product of their masses and inversely proportional to the square of the separation between the two objects." (It's sentences like this that give science a bad name.)

But the question I have is, "Why did he think it was a force of attraction?" Why didn't he think, "Every object in the Universe is repulsed by every other object . . . ." Couldn't the apple have been pushed down from above by the repulsion of the tree mass, or even by the repulsion of the heavenly bodies? He could even have proposed an invisible force called repulsion instead of an invisible force called gravity.

To take that thought further, perhaps I actually repulse the earth. (Hmm, that seems to be an unfortunate turn of phrase.) But that would mean the earth also repulses me. Then I'd be being held in place by all the repulsion of the other bodies in space which collectively overwhelm the earth's repulsion for me. If this were the case, wouldn't everything basically remain the same as far as our experience goes?

We hear a lot about how DNA makes proteins in living cells. These proteins are essential to cells in allowing other chemical reactions to occur. Some of the chemical reactions that proteins cause are the reactions that make DNA. We have come to think that life is DNA based. Why do we think DNA makes proteins instead of thinking that proteins make DNA? Further, how did one molecule come into existence, without the other molecule already being there, which couldn't have been there because the first wasn't there?

Don't misunderstand. I don't disagree with the findings of science. I just wonder how we started thinking in one direction instead of the other. I find thinking about thinking kind of weird.

There are really only six directions: forward, back, right, left, up and down. But that is actually only three dimensions. Well, I guess, if you are a Dr. Who fan, there is the time dimension too. But so far we haven't figured out how to maneuver about in time going straight at it. Anyway, if directional space doesn't work for you, and you can't move about in time, then there is really no place left to go.

Everyone seems to want to go someplace. If you're young, you want to go anyplace but home. Then you reach an age when all you want to do is go home. Some people want to go to Hawaii. Some people want to go fishing. Others just want to get away from all the hustle and bustle. Maybe I'm being redundant.

If we don't want to go anyplace, we often want to send someone else someplace. In that case, though, about the best you can do is send them, or go yourself, in the opposite direction. It's pretty limited.

Everyone complains about how rushed, busy, and strenuous it is in our modern world. The actual truth is that we in our modern world are extremely lazy, and the laziness is what causes all the hustle and bustle. For example, the streets are filled with traffic, and people are rushing through the airports to get home for holidays. However, the busy-ness isn't due to our activity, but to our repose. We are all sitting down in cars or planes, flying frantically across town or the continents, while reclining in comfortable, if crowded, seats.

Things would be far less frantic if we were far more strenuous. There would be less bustle if there was more activity. If we were compelled to walk wherever we wanted to go, we would be more active and less harried. But whether we are moving actively or lazily, we are limited as to where we can go by our material world.

When the world was a little less crowded, people could sometimes escape societal problems by running away. Noah did it, although he had a little help from a friend. Abraham did it and took a whole nation with him. The early American pilgrims, and immigrants ever since, were able to escape to a new continent when things turned ugly at home.

The problem seems to be that there doesn't seem to be any place to run to anymore. I guess we'll just have to sit tight and fix what we have.

## NO PLACE LEFT TO GO

Everyone wants their way made clear.
Should we leave or just stay here?
Where can we find the warmth of the sun?
Just waiting to hear from the anointed one.
Like with Prophets of old, we'd all like to know,
In truth, we'd all like to be told.
But there's really no place left to go.

In days of old, things weren't the same.
Prophets spoke then. Remember their names?
There was always someone to give us fair warning
To leave for the wilderness first thing in the morning.
They always knew where to go.
They had always been told
By a Prophet of old.

Everyone's going as fast as they can.
Most of us merely trying to stay alive.
I know I'm in a hurry to run away from me;
Always a day away from where I want to be.
Just once, I'd like to be told,
By someone who knows,
Which way to go.

I wonder if it will ever change.
Will there always be someplace strange?
Someplace to run, someplace to hide,
A safe place in which to abide?
I'd like to be told by a Prophet who knows.
Will it always be so,
That there's someplace to go?

Everyone I speak to is ready to leave.
The end is coming they really believe.
They're tired of the struggle, tired of the race,
Tired of the lies, and tired of the place.
But they don't know which way to go
Without being told by someone who knows
Like a Prophet of old.

# 16

# SEE HERE MY BROKEN BLADE

*"To try to be brave is to be brave."*
George MacDonald

I always hesitate to write about psychology. I might be opening up a can of worms that could take a very bad turn towards self-incrimination. Normally, being a parasitologist, a can of worms is right up my alley. However, psychology is very different from parasitology in that things are never what they seem. I mean, a worm is a worm. But even paranoid authors have enemies. That is why, now that I have made it clear that I am psychologically normal, I am telling you right up front that I refuse to answer any questions about this chapter on the grounds that it might incriminate me.

I have been so busy trying to save the world the last year or two, with absolutely no success, that I have fallen way behind on goofing off. My wife and I have different views on the value of goofing off, so I am always on the lookout for scientific data to support my position. My position, of course, is that it's too late to save the world, and I can do a better job of not saving it if I am rested.

So I've been catching up on my summer reading. Normally, I don't read science stuff in the summertime because I am sick of it after a school year of teaching and doing research. Then it dawned on me that I also don't read science in the winter because I am too busy doing research and teaching. So, right after I finished the Calvin and Hobbs book, I turned to a stack of papers on my desk

Taylor's Manifest Anxiety Scale is a test that has been used since the early 1950's to measure how much anxiety a person is experiencing. It's pretty simple actually. A person is asked to consider thoughts about themselves such as, "I often worry that something bad will happen." If you say, "that's me", it adds to your anxiety score. If you say, "not me", it subtracts from it.

My problem with these tests is that I never like the alternatives given. I would rather have a range of more honest answers like, "It always has!" or, "Only on days that end in Y!" However, Taylor has apparently been giving these tests to kids for over fifty years. The paper I read compared the kids of 1948 to kids of 1989. Here are some results.

| I wake up fresh and rested. | 1948 – 74.6% | 1989 – 31.3% |
| I work under a great deal of tension. | 1948 – 16.2% | 1989 – 41.6% |
| Life is a strain for me. | 1948 – 9.5% | 1989 – 35% |
| I have things to worry about. | 1948 – 22.6% | 1989 – 55.2% |
| I am afraid of losing my mind. | 1948 – 4.1% | 1989 -- 23.4% |

Now, it's obvious that, for the last forty years, our children have been growing more anxious. Other data suggests this trend has continued into the present. For example, another psychological test (the Rotter Internal-External Locus of Control Scale) has shown an 80% rise in external control of children and young people into their twenties from the 1950's to 2002.

Studies indicate that depression and anxiety increase in people when they feel as if they do not have personal control. In today's world, children under 65 have very little personal control.

As significant as this may appear, and as concerning as this is for our nations young people, I found it even more disturbing that I do not awake fresh and rested, work under a great deal of tension, experience significant strain much of the time, have more than my fair share of things to worry about, and have been accused of losing my mind. But that is just my enemies talking.

Forget the kids! I am suffering from anxiety. So, psychologically, I conclude that anxiety and depression are the new normal. Hence, I am normal. No more questions . . .

So apparently, anxiety is now the new problem in the world. It seems like this could be cured with a dose of courage. I guess courage is old-fashioned. It is a fascinating abstraction though. Courage is almost a contradiction in terms. I think it generally means a strong desire to live, in the form of a readiness to die. A soldier, surrounded by enemies, may only survive if he is fierce and careless about dying in a desperate attempt to live. If he does not fight, he is a coward and not courageous. He can't simply await death for that is suicide. He must fight for his life with a furious indifference to keeping it.

Many cultures see this differently. Christian courage seems to be a disdain of death. The Asian, of the past anyway, seems to have a disdain for life. The suicide dies for death. The hero dies for life.

I have difficulty distinguishing courage from faith. I suppose both are abstract concepts we tend to apply to the actions of others, seldom to ourselves. I think we often use both terms to describe behavior that we can't quite comprehend. I suspect most men wonder if they possess either quality: courage or faith. On the other hand, apparently, almost everyone is sure they have anxiety.

It seems almost mystical.  I think the best description I have ever read of courage is, "For whosoever will save his life shall lose it: but whosoever will lose his life for my sake, the same shall save it."

## SEE HERE MY BROKEN BLADE

I cannot speak, yet have so much to say.
I cough and choke on lungs of clay.
I have seen all the life I've lived.
I've given all that I can give.

> See here my broken blade,
> Used to shear her auburn braid,
> That she might not be betrayed.

Four things change like the season;
Time and knowledge, rhyme and reason.
You are young and think you know
How the way of your world will go.

> See hear my dented helm,
> Used to protect my queen and realm,
> Until I was overwhelmed.

I was hers, and she was mine.
Children arrived over time.
We claimed the woods and the land,
And all prospered from our hands.

> See here my shattered shield
> With blood and mud now congealed.
> She escaped because I did not yield.

Now, at last, I can hide my face.
This cloak of darkness I embrace.
I have withstood the enemy's blows.
I have conquered demon foes.

> See there's the gauntlet cold.
> It served me well, true, and bold.
> Now it and I turn to mold.

Though fleeting moments, I have endured,
Of this much I am sure,
My heart swears it's not been wrong.
I long to see you, it's been so long.

> See the lance shattered in two
> In spite of all that I could do.
> Each in the end receives his due.

Hand in hand, we stood our ground
Surrounded by baying hounds.
Even yet, with passion spent,
Until our very lives were rent.

> See the offering that I bring.
> A new world born in suffering.
> My pledge to thee is all I sing.

I once was trapped in life's parade.
It all seems now a promenade.
Seeming a king without a crown.
In retrospect, I seem a clown.

> I have stood where no man goes.
> I now know what no man knows.
> And I must go where all men go.

# 17

# SEEKERS ALL

*"As in all sweetest music, a tinge of sadness was in every note. Nor do we know how much of the pleasures even of life we owe to the intermingled sorrows. Joy cannot unfold the deepest truths, although deepest truth must be deepest joy."*
*George MacDonald*

Have you ever wondered why we call numbers that are divisible by two "even" and all the others "odd"? What makes things like one, three, five, and seven any more odd than two, four, six, or eight that we appreciate? There are just as many odd numbers as there are even numbers, so it's not like odd numbers are some kind of minority.

I think it is just as odd to be even, but I suppose that may be because I am odd. The natural world appears more odd than even. I mean, perfect symmetry is a rare thing in nature. Even when there is a pattern, it is often a little uneven. Doesn't that make it odd? In a world where it is improper to call odd things odd, why are we still using hurtful terminology to talk about numbers?

This is just another sign of the coarsening of American society. Even math has become polarized with hurtful and ugly things being said on both sides. In fact, the odd and even thing is on a par with terms like "he" and "she" or "male" and "female". It's all sexist. I didn't ask to be born with an odd chromosome. It's a birth defect, and polite people shouldn't make fun.

There might be a better system. The Greeks had a promising idea. They didn't have calculators so used marbles instead. Yes, the Greeks claim to have invented marbles because a taw and three aggie ducks were found in the ruins of Pompeii. If you don't have all your marbles, though, you can use little pebbles to follow this along.

Starting with one pebble and adding two pebbles gets you three. Three pebbles can be stacked into a triangle. Since you added two to the first pebble, you now add three, which is the next number in sequence, to the three pebbles and you will get six pebbles. Six pebbles also stack into a triangle. The next number in sequence is four and if you add four to six you get ten which can also be stacked as a triangle along with the three and the sixj. Continuing this pattern of adding the next number of pebbles in sequence will always give you a number of pebbles that can be stacked into a triangle. Therefore 1, 3, 6, 10, 15, 21, etc. would all be "triangular numbers".

Now, what if you start with one pebble, and instead of adding two pebbles, you start by adding three pebbles? That would yield four pebbles, and four cannot be arranged into a triangle. Four must be arranged into a square. From here, add three pebbles again, plus two more for a total of five, and you get nine which is also arranged as a square. If you add five pebbles plus two, seven, to nine you get sixteen. If you continue this pattern of adding the previous number of pebbles, plus two, you will always get a number of pebbles that can be arranged into a square. Hence these are called "square numbers", but are not to be confused with squared numbers. The road goes on forever, and the party never ends.

An interesting feature of this arrangement is that if you add two triangle numbers together, you always get a square number. And, of course, all square numbers can be divided into at least two triangles.

Looking at numbers this way completely eliminates the concept of odd and even numbers and replaces it with triangle and square numbers. Since some triangle numbers are even and some square numbers are odd, there will no longer be a stigma attached to either. This type of arithmetic is called "figurative arithmetic" which you probably thought was something else entirely, you sexist.

A distinct advantage to triangle and square numbers is you get away from the whole idea of opposites which is so hurtful. So, adopting this old Greek method would go a long ways towards making mathematics more politically correct. It's something that has long been needed, but math is rather rigid and old fashioned.

The problem with opposites is that they cannot exist without the opposite concept. That's why they are called opposites. They are defined by each other. Reflections are a special form of opposite. You see the same image as the original, except it is flipped. If you actually move your right hand, it is on your right in the image. But if you face the same direction as you are in the mirror, it is your left hand.

What is interesting, though, is that if you look at a reflection of a reflection it will also be flipped so that the reflection of the reflection will look just like the original. This is also true if you look at a reflection of a reflection of a reflection. It goes on forever.

I wonder how much of this can be found in other concepts that seem opposite. When I observe a painting, do I see what the painter saw or is it somehow flipped? If I were to try to paint it back to the painter, would he see it flipped and think it was just like his painting? This is just a metaphor since I am sure that any painting I painted of a painters painting would be insulting to the painter.

But there is another way that we are mirror images of each other. Painters need someone to see their painting. And people need paintings to see. The painter needs someone to look at his painting to see if it has communicated what he wants to communicate. He needs his effort verified, or else corrected.

However, people need to see what other people see so they can tell if what they are seeing is real or not. If you see a painting and do not think it is "good", it causes you to question whether you have an accurate view of the world. Perhaps the painter is showing you something you had not seen or considered before.

Teachers need students, perhaps even more than students need teachers. The curious, hungry student will learn without a teacher. A driven, curious teacher can do nothing without a student. A banker needs the depositor as much as the lender needs the bank. The builder needs buildings to build and buildings need builders.

It's almost as if an identical thing needs another identical thing to define it. Everything needs everything. Doesn't everybody need somebody?

## SEEKERS ALL

Seekers in the night are drawn to the flame.
They begin to arrive as darkness falls.
With many shapes and numerous names,
They answer some still unknown call.
Like pilgrims seeking a mystic source,
Joined hands in ecstasy,
They follow a faint and ethereal course
And on arrival, they bend the knee.

The air has power and to spare
That spills and fills the empty hall.
And patient eyes watch from below.
Brooding silence settles like a pall.
The wanderers wonder and await the sound,
The wind, the power of exultation.
They seek what others have already found
In the air, exhilaration.

Those above watch those below
With impatient majesty.
Watch the others as if they know
How it feels to be free.
Though they harness power of air
And seem to be the source of light,
The truth is they do not dare
To remain out of the pilgrims' sight.

At last the sound explodes the walls,
Pouring forth in harmony.
And though the pilgrims sit enthralled,
The magicians themselves not really free.
For as the pilgrim seeks the sound,
The sound must have the listening ear.
And as those above must be seen
By those who worship sound and cheer.

When at last the wind dies away,
Silent brooding and stillness falls.
The stars fade to a new day
That puts on a trash-strewn shawl.
The seekers have gone to whence they came.
The magicians have quit the stage.
Each of them with secret shame
Has returned to his separate cage.

# TEAR DROPS TURNED TO RHYME
*"No story ever really ends, and I think I know why."*
*George MacDonald*

I have been thinking about science and suffering.

I became a scientist, in part, because I wanted to help people. The other part was avoiding the draft. But then I got drafted anyway. Now I find myself administering exams and writing columns, both of which may cause excruciating suffering for some people.

Many scientists seem to worry about suffering. However, the scientists who worry the most are people who have never seemed to suffer very much. Richard Dawkins, Victor Stenger, and other atheist scientists are greatly exercised about suffering although they, themselves, have university educations and live rather extravagant and indulgent life styles.

I'm not entirely sure what suffering is. Presumably death qualifies, although I really don't know, not having tried it yet. I suppose pain causes suffering, but pain exists on some kind of a continuum. To what degree do we suffer? Does an athlete who "suffers a loss" experience the same suffering as a family who "suffers a loss"? If I skip meals in an effort to lose weight because I suffer from being overweight, am I still suffering from hunger?

I meet many people who are deathly afraid of being stung by a bee. I'm not sure I would call getting a bee sting suffering. What about having only one shirt? Is that suffering? I guess that depends on how often you are able to do the laundry. What if you don't own a car and have to walk? That might depend on how far. What if you have to live in a one-room shack? Well, that depends on who your roommate is, doesn't it?

Interestingly, many people who live under poor conditions don't act as if they are suffering at all. They laugh, love, play, get married, have children, and seem to have a good, old time anyway. In fact, they don't seem to worry too much about suffering. They often seem happier than the rest of us.

I'm not sure why scientists worry so much about suffering. Suffering isn't a science, although my wife thinks I have turned it into such. She just doesn't understand how hard it is to be a professional windbag. Sometimes my back kills me from standing through all those lectures. I wonder if being boring counts as a disability. Do my students suffer?

If science is so interested in alleviating human suffering, why do almost all the things scientists discover to lessen suffering, end up causing more suffering? After WW II, world scientists decided to eliminate Malaria. Things fell apart politically after a few years, and they gave up. By then, though, more people were dying from starvation because of overpopulation. Overpopulation was partly a result of more people surviving Malaria.

Scientists developed antibiotics to fight disease. But through the use of antibiotics, disease organisms became resistant to antibiotics. Wouldn't it be ironic if we relied on pesticides to grow more food to feed the world, but the pesticides killed the bees, so we couldn't grow more food? No!

There are those who seem to be concerned about who is responsible for suffering. Interestingly, some people think it is God's fault for allowing suffering to happen. Shouldn't they blame Satan for causing it to happen? If I fail to stop an accident, am I guilty of the suffering the accident causes? If a scientist fails to find a cure for cancer, are all the cancer cases the fault of scientists?

Well, the truth is I became a scientist for several reasons. One was I discovered that I couldn't get a job with a degree in English literature. In addition, I liked working with things more than with people. I think I like humanity, as a whole, more than I like a lot of individual humans. Actually, I just have a lot of fun playing around with bugs.

Well, thankfully, I don't have to assign blame for suffering. If, like Dawkins, I never alleviate any suffering, I can always blame God. Somehow things never really seem to change.

**TEAR DROPS TURNED TO RHYME**
On a morning he knelt praying,
A heavenly ordained date
That would affect hungry souls.
Millions lay in wait,
But lust for gold and power
Interfered with the plan.
Wicked men and evil design
At last destroyed the man.

It started in New York
When the tear drops turned to rhyme.
Then followed to Palmyra
With the sound of children crying.

See it there, shining in the sun,
Fist Temple in a thousand years.
Rushing wind like days of old
Descends upon the seers.
But the curse of Kirtland
Lays upon the land.
Murders and mobs
Allowed by God's hand.,

It started in Ohio
When the tear drops turned to rhyme.
Then followed to Kirtland
With the sound of children crying.

Listen stranger, you're not our kind,
And you know we were first.
We still enjoy drinking,
And you enjoy thirst.
We aren't the friendly neighbors
On which you laid your plans.
So move along and no one's hurt,
This ain't your promised land.

It started in Missouri
When the tear drops turned to rhyme.
Then followed to Illinois
With the sound of children crying.

Look away cross that river.
We settled for the rain,
Swamps, work, and bitter toil
To try to grow some grain.

The world was filled with trouble
And again we were deceived.
In the dead and dread of winter,
Again we had to flee.

It started in Nauvoo
When the tear drops turned to rhyme.
Then followed to Utah
With the sound of children crying.

We'd like to have a home,
A place where we can stay.
But we never knew there was
Such a price to pay.
God can tell us plain
Exactly what to do.
But he doesn't make it clear -
The result when we are through.

It started in Palmyra
When the tear drops turned to rhyme.
Then followed throughout our quest
With the sound of children crying.

# 19

# THE BUSINESS PLAN

*"The man who was lord of fate,*
*Born in an ox's stall,*
*Was great because He was much too great*
*To care about greatness at all."*
*George MacDonald*

Ben Franklin must have had too much time on his hands. I mean, who goes out in the evening to pour a quarter cup of oil on a pond to see if the oil will calm the wind-chopped water? If he were to do something silly like that today, he would probably miss an entire episode of American Idol! Does missing reality television, so one can experience the real world, seem odd?

Today we know that oil doesn't mix with water. It forms a film just two molecules thick that spreads across the water's surface and acts a little like a skin covering that stops the wind from rippling the surface. Of course, in today's world, Franklin would be sued by the environmental protection agency for creating an oil spill. Another question: does oil on the water pollute the water or minimize erosion? I guess I could look into that if I didn't have to catch the rerun of the Biggest Loser . . .

Then there was all that time Franklin wasted fooling around with the Leyden jar. The Leyden jar is a glass jar with an inner and outer metal covering. A metal rod, inserted into the jar through a wooden stopper, is connected to the inner coat by a metal chain. It was invented at the University of Leiden in the Netherlands by a "natural philosopher" (that's what they called time-wasters in those days) named Pieter van Musschenbroek back in 1746.

Franklin actually built several Leyden jars, apparently for the sole purpose of shocking his dinner guests for after-dinner entertainment. They would have all been just as happy to have gotten a tweet about the experience, but he couldn't describe it in just 140 characters. Franklin even went so far as to hypothesize that there were two, separate charges stored in the jars inner and outer metal layers. He called these positive and negative charges. These, he said, were separated by the glass until something, or someone, created a pathway for the two to flow together. He found providing dinner guests, as a conduit, shockingly entertaining.

It isn't clear whether or not he actually wasted time flying a kite in a thunderstorm. But it was a Leyden jar that sparked an interest in doing so. That was the jar over which the infamous key supposedly hung. And it was his thinking and tinkering with electricity that led to the invention of the lightening rod.

Then there was the whole blowing-on-the-thermometer-bulb-that-had-cloth-wrapped-around-it-that-was-moistened-with-chloroform event. He and a friend did this just to see if he could drop the temperature as low as 7° F on a hot summer day. I don't know who was running the print shop, but I'll bet he never watched an entire Rockies game.

When he wasn't wasting time on reality, he was partying. One of the social organizations that he founded was the American Philosophical Society. He did this in 1743 even before the Leiden jar had been invented. The society was meant to be a place for prodigious time-wasters to discuss ways for wasting time.

He completely missed the Telluride Bluegrass Festival and Country Jam because he was tracking storms, investigating water spouts and whirlwinds, charting and naming the Gulf Stream, identifying electrical conductors, and inventing urinary catheters, bifocals, more efficient stoves, and catamaran hulls. He was such a prodigious waster of time that he actually won an award. In 1753 he was awarded the Copley Medal by the Royal Society of London. The Nobel Prize had not yet been established, so this was the highest scientific award of his day. In 1756 he was elected a member of the Royal Society of London, a group of similar, social misfits who never even tried to play "Battlefield Bad Company 2".

Benjamin Franklin was one of the most accomplished scientists of his day, a fact that has been mostly forgotten. The esteem, by which he held influence later in life, was in great part due to his scientific reputation. I wonder how a person like Franklin would fare in today's world. Is there any place left for a unique person, with many interests and activities, that aren't just like everyone else's?

**THE BUSINESS PLAN**
To stand at the chiasmus of two eternities:
Past and future, and arriving at the boundaries
In just the nick of time.
Notching and noting it on my walking stick,

Having made good time with little traffic,
And it being a gentle climb.

I hope you will pardon some detail obscurity.
There are more secrets in mine than some men's stories.
Not due to any crimes,
Just inseparable from its very nature
The warp and woof of its very queer features
And peculiar, odd pastimes.

How can I explain the search for a missing hound?
Stumbling along blindly to follow the sound?
I am still hot on the trail.
Or the assistance I've rendered to the bright, rising moon?
Though never to influence whether it was late or soon,
Still, my presence did avail.

The self-appointed inspector of sun, rain, and wind,
Patrolling sidewalks without complaint to chagrin.
I did my duty faithfully
For a very long time, I may say without boasting.
I did what I did without pay, never coasting.
Duty with humility . . .

But it, then, became increasingly evident
That I lacked something in the community's judgement.
No room in the inn it seems.
I would not be admitted as a town officer,
Nor given allowance for a place secure,
Just self-responsibility.

So I calculated a small business plan.
I counted and estimated before I began.
Experience was helpful.
To manufacture dreams from sun, rain, and wind,
And market boldly across continents,
Bright shiny beads and baubles.

I felt I could acquire wealth if I tried.
Until God helped my ambition die,

And I desired to just be still.
The one who desires to serve the son
Will become God's chosen one.
Let Him make you what He will.

# 20

# THE DOCK

*"I would rather be what God chose to make me than
the most glorious creature that I could think of; for
to have been thought about, born in God's thought,
and then made by God, is the dearest, grandest and
most precious thing in all thinking."*
*George MacDonald*

People often have favorite colors, or words, or names, or songs, or cars. But they seldom seem to have favorite numbers. In fact, I tested this observation with a class and found that most people did not think numbers were important enough to be given favored status. If they had such a number, it was considered a lucky number because of its relationship to some emotional event or object.

I think some of my favorite numbers are eight, sixteen, twenty-four, forty, 6/5th's and 1/3. It is because these numbers represent proportions in the world of Lego engineering. As such, they can be used to build significant objects and to engage student minds. Most expert builders with Legos are under twelve years old and are unaware that Legos are constructed with unique engineering standards.

If you are new to the world of Lego construction, you need to know some terminology so you can follow the discussion. The rounded protrusions that fit into the bottom of other pieces are called studs. Lego pieces with a single row of studs are called beams. If there are two rows of studs, they are called bricks; and if there are more than two rows, they are called plates. The bricks, beams and plates are further distinguished by the number of studs; there being eight stud bricks, ten stud bricks, four stud beams, etc. That's enough for now.

Lego bricks have a vertical unit that is 6/5th's the horizontal unit. That is, a stack of five bricks is the same height as the length of a six stud beam. This is helpful to know when you are trying to build a complex structure. If you build structures with vertical heights equal to horizontal lengths, you can utilize braces. Plates, on the other hand, are 1/3rd the height of a brick or beam. So it takes three plates to be the same height as one brick.

Lego gears are always differentiated by their number of teeth. They come in four sizes: eight teeth, sixteen teeth, twenty-four teeth and forty teeth.  Notice that these numbers are related to each other: they are all divisible by eight.  When an eight-tooth gear, intermeshed with a twenty-four tooth gear, rotates one complete turn, it will turn the twenty-four tooth gear one third turn.  That will cause a 3 to 1 reduction in speed and gain in power.

If an eight-tooth gear intermeshes with a sixteen-tooth gear, every rotation of the eight-tooth gear will cause the sixteen-tooth gear to turn one-half time.  That is a two to one decrease in speed and increase in power.  This process is exactly what happens on your mountain bike, except the gears are meshed by means of a chain.

That raises an interesting question about gear chains.  What if you hooked up a series of gears in a row?  If you set up two three-to-one gears in a row, the final power decrease would be a nine to one loss. If you were to set up a three-to-one gear, followed by a five-to-one gear, the decrease in speed would be fifteen-to-one.

There are other interesting numbers used in Legos, though I'm not sure they are as important as the one's I've mentioned.  According to "The Ultimate Lego Book" (DK Publishing, 1999) two eight-stud-bricks can be combined in twenty-four different ways.  Three eight-stud-bricks can be combined in 1,060 ways.  Six eight-stud-bricks can be combined in 102,981,500 ways.  By the way, this book would be an inspiration to any budding Lego builder.

Building is perhaps one of the most-educational things humans can do.  I cannot know what a person understands, but I can gauge his understanding his behavior.  Building is one of the most complex behaviors we humans engage in; and if something is not well built, it is visible to all.  "For which of you, intending to build a tower, sitteth not down first, and counteth the cost, whether he have sufficient to finish it? Lest haply, after he hath laid the foundation, and is not able to finish it, all that behold it begin to mock him, saying, this man began to build, and was not able to finish."  Luke 14:28

Perhaps the other number that is important to the Lego world is the 1932.  That's when Kirk Christiansen of Denmark began making toy trucks to enhance his carpentry business because of the great depression.  He filled his trucks with little wooden bricks and found people returning to buy more bricks.  Lego was born.  Maybe 1932 is my favorite number!

## THE DOCK

They say on a distant shore there is a land where promise waits.
There, a million starving souls arrive each day at the gate.
But none can pass, they say, without the Kings consent,
And no man seems to know just where the King went.
Only those can enter who have attempted purity
And possess the courage to be free.

They say that the snow in the mountains melts and flows beneath the sand
In a great underground river that is inaccessible to man.
Across the sand can be seen an oasis, or so I've been told,
Where the sun and thirst are forgotten. A mirage that destroys lost souls.
Only the dead can go there. Only the souls that are free.
Only the part of a man that has entered eternity.

While a man is there, he must labor and build with his own hands.
Using the tools of his trade, the dock is where he must stand.
They say he will stand at the dock and hear the charges called,
Ashamed of the shoddy construction built with his hammer and awl.
Each man in his life is driven, whether by words or a riveters' gun,
To build with the tools he is given, until his own cell is done.

Height, width, and depth to be measured, whether by God or by man,
Honest labor and heart will be reckoned, then lifted by the Saviors hand.
On a night as I lay sleeping, I dreamed like a thirsty soul
For just a taste of that water that they say refreshes lost souls.
To be freed from the stocks of mortality,
And to be what I promised to be.

The second is the sign of marriage. The fifth is the sign of life.
The third is the sign of eternity, the infinite blessing of life.
You know full well I loved you. The words I just couldn't say.
Working was all that I knew; the other just not my way.
Imprisoned alone in the end, after all that I've done,
My only way out of the desert is to be freed by the life of the Son.

Each man in his life is driven, whether by words or a riveters' gun,
To build with the tools he is given until his own cell is done.

# 21

# THE GUIDE

*"It is by loving and not by being loved that one
can come nearest to the soul of another."*
*George MacDonald*

People often tell others where to go.  People sometimes give their computers similar instructions.  I've noticed that people never offer to guide me where they tell me to go.  Or maybe telling me where to go is a form of instruction on how to get there.  This "knowing where to go" is not easy.

However, people and computers are very different.  First, the computer wants to know, in advance, how far to go.  But often we aren't too sure how far it is to where we would like the other person to go.  The other difference is that, when we tell a computer to do something, it usually does it.  People never go where I tell them.

Once I was teaching a young man to program a computer to draw a geometric figure on the computer screen.  He learned to tell the computer pen to "go forward 100 steps", steps being a relative term to what a computer does.

Then he told the computer to "turn right" without stipulating how far right to turn.  Now, if a person or computer turns right far enough, they will end up going in the same direction as before.  Likewise, humans often tell each other to "turn their lives around", apparently not understanding that this could be fruitless.  It'd be like doing the Hokey-Pokey.

In the case of a square, computers need to be told to turn 90 degrees.  However, once these two basic instructions have been quantified, you have the makings of a square.  Simply tell the computer to repeat those instructions four times to create a square.  Can you imagine a computer reading the instructions to shampoo its hair?  The instructions to "apply, lather, rinse and repeat" never end.  A person would drown if led by a computer.

The young man's final program looked something like this: Repeat 4 (Forward 100 Right 90).  From there, triangles were a cinch.  Merely change the repeats to "three" and the distances and angles to appropriate values.

My student had to stand up and walk in a circle to figure out how to program a circle. He started by having the cursor move one step forward and then turn one degree. Then he added the command to repeat that process several times. This created a nice arc but not a complete circle. He kept increasing the number of repeat commands to create larger arcs. Suddenly he stopped and exclaimed in an excited voice, "That's why they call it a three-sixty!" He then proceeded to close the circle by writing the program: Repeat 360 (forward 1 right 1)

This youth was already an accomplished skier and knew how to jump off a mogul and turn completely around. He knew it was called a 360, but he didn't know why. Isn't it interesting that his experience in doing something physical, like walking in a circle and skiing, enabled his breakthrough in being able to do something intellectual? The body teaches the brain before the brain can instruct the body.

Notice that the computer already knew what a degree was and that it took 360 of them to make a circle. Humans had told the computer that. But how did a human first figure out an exact 90 degree angle before there were computers?

Draw a circle using a compass. Draw a second circle of the same radius by placing the compass point on the edge of the first circle. The two interlocking circles will create an "eye-shaped" figure where they intersect. Today we call these Venn diagrams. In the old days, they called this "eye shape" a fish bladder or Vesica Pisces.

Now draw a straight line between the center points of the two circles. Finally, connect the corners of the eye shape to each other with a straight line. Where the two straight lines intersect, they form an exact ninety-degree angle, necessary for constructing squares.

How do you suppose some human taught that to a computer? The answer, of course, is that the computer doesn't know this at all. Only humans can figure stuff like that out by walking in circles. Computers could never figure it out because they can't walk in circles.

So if you plan to tell a computer or a person where to go, you might want to consider specific instructions. Better yet, show them the way.

## THE GUIDE

Once upon a black stormy night
Beside the fire, alone and cold,
I pondered, upon ancient signs,
The hidden stars I knew of old.
The candle flickered in the air,

The fire just an embers glow,
I must have dozed in my easy chair
For, what I had dreamed, I do not know.

For a moment, it seemed as though
God's breath was on my brow,
A lost warmth within my feet,
And life seemed enough for now.
A moonless night obscured my sight.
But it was just the wind, not a friend,
That awakened me in the night
So far away from nighttime's end.

In the dark, I lay in half-slumber
With no one else by my side.
And remembered things as they were
Before my only love had died.
With just the sound of the gentle breeze
To take away my sleep and dreams,
Filtering through the leaves and trees,
It rippled like a moving stream

Through the dark, I stared in wonder
At shapes and shadows in the night.
The wind visible in the quiet
Restless, moving, in its flight.
And I thought I heard wind whisper,
Though I could not be sure,
If it uttered questions or spoke answers
Whether warning or allure.

There is sound, within the quiet,
That can set the heart in fear.
There is quiet, within the sound,
Perceived by the listening ear.
Uncertain, sad, unsettling whispers
Speaking of eternity
Uncertain, sad, unspoken longings
To, at last, be set free.

Breezes blow, as soft as smoke,
Like entrails of the stars.
Time passes, swift and slow,
And leaves behind blood and scars.
Memories flood my dreams,
Or my dreams are memories.
But, in the night, I recall
How much my love loved the trees.

What comfort this sensation gives
Like the spirit of Pentecost.
Awakened by the still small voice
That assures all is not lost.
For there, in the darkest night,
My love blew by me like the wind.
She grasped my hand in passing
And led me on to the end.

# 22

# THE MORTAL GARDEN

*"Seeing is not believing - it is only seeing."*
*George MacDonald*

There is a double standard, used by many scientists, to curry favor and research grants. For example, physics has portrayed itself as a great beneficiary of mankind, all the while providing us with the atomic bomb as well as electricity. Did you know that almost 4000 people a year are killed by electricity?

Chemists like to think of themselves as benefactors of mankind because they can make plastic and pesticides. Now we are drowning in both.

For some reason, the zoologists get pegged with all the dangerous animals from lions, tigers, and bears, to rattle snakes, scorpions and wasps. Does anyone ever mention the bees and beetles that pollinate our crops or the dung beetles that help decompose dung?

Then there are the poor, little, peace-loving botanists who just grow flowers, food, perfumes and hallucinatory drugs. They are not so innocent, let me tell you.

Take Aconite, for example. Well, no, don't take it literally. Aconite is known as Wolfs bane or monkshood, and eating it lowers the blood pressure and stops the heart. The whole plant is highly toxic, but the root resembles a horseradish root or a pale carrot, so I guess vegetarians get confused

Everyone covers up their wall sockets when children are born, but nearly 70,000 people a year are poisoned by plants. You think because you don't eat horseradish you're safe? Well, that Philodendron can cause nausea, severe abdominal pain, and serious allergic reactions. Dieffenbachia can inflame and paralyze the vocal chords leaving a person unable to speak. It can cause much worse symptoms, so don't get any ideas. I could still write.

Sure. Tell me about the kind, peaceful botanists who harbor such things as Castor Beans, the plant that gives us ricin. It is now obvious that my mother was trying to kill me when she gave me castor oil as a child. Supposedly a laxative? Oh sure! This is the same stuff the KGB used to kill Giorgi Markov, the BBC Journalist, back in 1978.

Okay you vegetarians, what about Lathyrus sativus, or Grass Pea?

Sure, it is a good source of protein, but it also contains Beta-N-oxalyl-diamino propionic acid. You know, that's the neurotoxin that kills the nerve cells in the lower limbs and leaves a person paralyzed from the waist down. They say that if you soak it long enough in water, it is safe to eat. How long?!

It isn't an accident that Alfred Lord Tennyson wrote of the Yew tree, Taxus baccata, "The fibers net the dreamless head / Thy root s are wrapped about the bones." In fact, in 1990 a yew tree was uprooted during a storm and skeletons were found in its roots. The whole plant is poisonous, but the sweet tasting berry causes the biggest problem. It contains a poisonous seed, and eating the seed causes a drop in pulse rate and heart failure. Scientists seem a little unclear about the exact symptoms because often the first evidence of yew toxicosis is death.

Don't misunderstand me. I don't blame the botanists that these plants exist. They didn't create them. God did. It just gets a little tiresome hearing all about the healthy plants, garden clubs, and recipes for innocent Habanero Chili when there is a dark side that is hardly ever acknowledged.

I have nothing personal against botanists. Some of my best friends are botanists. I think botanists should have the civil liberties accorded to any minority group and should never be persecuted. I think it's kind of like a birth defect or something.

However, God planned a mortal existence that is fraught with peril. The garden was carefully planned to see what we would do with it. God knew that there had to be dangers and difficulties for us to grow. He planned the garden carefully.

### THE MORTAL GARDEN
From high overhead through my window
I watched the gardener grow
The crops he planted carefully.
Yet it seemed as though
He was tentative, perhaps afraid.
I did not understand why it should be so.

He walked as if among
A malignant and unseen force,
Savage beasts, or deadly snakes.
Fearfully he forged his course
As if seeking among the dead

For life's eternal source.

Safe within my room on high,
I wonder if what I see is true.
"Seeing is believing" I've heard it said.
I suppose that depends on the view.
But the gardener still moves with care
Whichever window I look through

He's preparing wine of forbidden fruit,
Planting sorrows that we sow,
And pruning thorns and thistles
Of the things we need to know.
The garden is delicious,
Warm, medicinal, and show.

I'm searching for the garden
To try and be what I can be.
I'm looking for the entrance
To see what I cannot see, and
To smell and touch the fruit
Of the forbidden tree.

For my window to the garden
Is never open all the way,
But I can scent the flowers
And the smell of broken clay.
I long to feel the sunshine
Of the gardens bright new day.

Is it quiet in the garden?
Are there voices from the past?
Or are they all forgotten
In the shadows that are cast?
Will I recall my windowed room
When the garden wall is passed?

Beautiful, deadly shadows
Sprinkle sunlit ways,
While dangerous lovely flowers

Create a golden haze.
With care and faith and courage
I will navigate these earthly days.

# 23

# THE PREACHER

*"Here there is no room for ambition. Ambition is
the desire to be above one's neighbor; and here there
is no possibility of comparison with one's neighbor: no
one knows what the white stone contains except the
man who receives it.... Relative worth is not only
unknown to the children of the Kingdom it is unknowable."*
*George MacDonald*

There are only four seasons of the year in which conditions are just right to cause me to become pensive: that is, engaged in deep, serious thought. It doesn't last long in the winter as I tend to fall asleep soon after becoming pensive. In the summer, I don't have much time for "pensivity". But in spring and fall, I can sometimes spend as much as several minutes engaged in deep, serious thought. It never lasts though. I always revert back to my normal, shallow, silly thought.

During a recent pensive fit, I remembered that "all truth can be prescribed in one, great, eternal round". I couldn't quite make out, before the fit subsided, whether it was an actual circle, an ellipse, or maybe a spiral. I know it was definitely circular. My passive fit happened as I contemplated my bees buzzing around their hive, the tulips and daffodils just barely sticking up through the ground, and that darned squirrel scampering in my wood pile while waiting for my tomatoes to come on.

I know none of this is any great thought or discovery in general. People have celebrated the seasons ever since there have been people and seasons, I'd guess. However, we don't usually think of science as being circular. Humans see science in a linear way. Each new discovery opens the door for the next new discovery and, with every new discovery, comes ever greater and greater understanding.

So how come it doesn't work that way? Science seems to grow upon itself, and our understanding of physical laws and nature has increased to an astounding degree. But the problems never seem to be solved. You'd think we would eventually know enough to solve something.

We make better weapons, but we still have wars. We know more about genetics, but we still take away peoples' civil rights in the name of protecting someone else's civil rights. We know more about the environment, but we agree less and less on how to care for it. We know more about cancer, while millions still die from Malaria and parasitic diseases.

We don't solve problems with better technology. We solve problems with knowledge. I think that may be exactly the problem. Machines, hardware, and software have no knowledge. Only humans have knowledge. Technology is simply the product of our knowledge. Since it is knowledge that solves problems, that means that only humans can solve problems.

The problem is that humans appear to be "the problem". It's not that humans are bad. It's just that humans are circular. Children come into the world without knowledge. They grow, learn, and increase in knowledge while alive. However, children are born into a world that their parents have already messed up while they were trying to fix it.

The result is that each new generation must fix the problems caused by the previous generation. Today's generation, therefore, needs different kinds of knowledge than their parents used to mess it up. By the time each generation has figured out the world of their lifetime, the world has become something else again. Each generation needs different knowledge, so they can mess the world up in their own way. There is no forward progress, just generations running in circles.

The people keep cycling. The science and knowledge keep growing anew. Of course, pensiveness occurs in cycles, too, because only people can be pensive. Who ever heard of a pensive cell phone? I wonder if animals are ever pensive. I thought I saw a pensive bee once, but it turned out to be just dying of old age. Hmmmm.

Are young people ever pensive? I can't remember. Are scientists, besides retired ones, ever pensive? It would seem like a pensive scientist might lead to further scientific knowledge even though the scientist, being human, is cycling and will be replaced by another generation of scientists. What do you call a science-column writer who has stopped being thoughtful? Ex-pensive?

Well, I am not the first to discover this. Ecclesiastes is one of 24 books of the Tanakh, or Hebrew Bible, that explores this same theme. Ecclesiastes is classified only as one of the writings. It is also a part of the Old Testament in the Christian church and is found among the canonical Wisdom Books in the Old Testament of most denominations of Christianity. The book is even popular among atheists.

Ecclesiastes is derived from a Greek word that means "gatherer".

However, the anonymous author referred to himself as "The Preacher", or some say "The Teacher".

Ecclesiastes has enriched the English language. There is "nothing new under the sun". There is "a time to be born and a time to die," and "vanity of vanities; all is vanity" all come from this book. During some of the most difficult times of the civil war in 1862, Lincoln quoted, "One generation passeth away, and another generation cometh: but the earth abideth for ever...." The book concludes, "Fear God, and keep his commandments; for that is the whole duty of everyone".

While the overall theme of the book seems somewhat dreary, declaring man's efforts vain, useless, fleeting, transitory, and meaningless, it also helps mankind to understand that our troubles, seemingly unique to us, are not. They have been experienced by all men and will probably be experienced by those yet to come. It pretty well dooms the idea of social progression but not of individual growth and understanding.

I think it is a little like singing the blues. For some odd reason, it makes you feel better after lamenting about how terrible it all is. If the following sounds familiar, it's because it is mostly lifted straight from the book of Ecclesiastes with care given only to rhyme and meter. Fear God! Keep his commandments, and enjoy life.

**THE PREACHER**
The things that have been
Are the things that will be.
Things that were seen
Are the things we will see.
The deeds that are done
Have already been done.
There are no new things
Under the sun.

Each day the sun rises.
Each day it goes down.
Tomorrow until forever,
One eternal round.
The wind blows to the north,
Then blows to the south.
The wind blows everywhere,
And whirleth about.

The eye is not filled with seeing,
Nor darkness when we sleep.
And tears are not diminished,
No matter how much we weep.
Ears are not filled with hearing
At least until we die.
Words are endless on our tongues
Whether truth or lie.

The river runs to the sea,
But the sea is not filled.
The tides rush on and on,
And it seems forever will.
Is there anything whereof
It can be said,
Anything new is to come,
After we're dead?

# 24

# THE TEACHER

*"They think, if they think at all, that it is life,*
*strong in them, that makes them forget death;*
*whereas, in truth, it is death, strong in them,*
*that makes them forget life."*
*George MacDonald*

Imagine a narrow beach with high ragged cliffs, a grey sky, cold and whipping winds, and pounding breakers from the North Sea. A ragged band of men dressed in skins and furs, wading ashore, wild hair and beards blowing in the wind, sweeping out of the cold skies. The men carry swords and spears and are wary as they enter a new land that is untouched and untrodden. The first band finds a way up the cliffs face, and there they find open moors and, further off, dense forests. In the days that follow, more men wade ashore from the waiting ship and small parties fan out to explore their new home.

The moors seem endless, wild and featureless. The inland forests are pristine and, at times, impenetrable. But these men have known the featureless sea, and they come from stock who are at home in the forests. They plunge ahead in their explorations. In the midst of large expanses of endless trees, occasionally there were clearings. In these, the sunlight seemed more intense and striking to them in contrast to the expanses of dark and ancient forests. Their name for such a clearing was "lea", derived from "leocht", their word for light.

The plains, in their sameness, and the forests, with their obstruction of sight, presented challenges for these sea-faring men. As they ranged farther from their beach establishments, they sometimes became disoriented, and finding their way was difficult. But there is an advantage to the land that the sea does not possess. Their passing leaves a mark on the land: a footprint, a broken branch, a scratch on a rock. By carefully following the tracks that they or their fellow explorers had left behind, they could find their way again to distant places, or home.

Of course, as they followed one another's footprints through the forests and meadows, the trail soon became clear, then worn, and eventually a depressed path, almost a furrow, marking the way to travel. Their word for footprint, track, and furrow was "leis", obviously taken from "lea", the word they used for the forest clearing. The two are connected because a track provides information, similar to shedding light on a subject. They also had a special word that meant to follow the track, to benefit from the knowledge of those that came before, to memorize the way, and to study out the way to go. They called the process "leornian", or "lernen", the word that became our modern word "learn."

As the paths became established, one who had traversed the way many times could describe the paths to someone who had not been there. Of course, the traveler wouldn't describe every step. That would be too confusing. But the experienced traveler could describe the major landmarks, tell where to turn at branches in the trail, and estimate about how far to go before someone would expect to see the next landmark.

Eventually the path became many paths throughout the land. Some were worn into deep furrows; some were barely visible except to the sharpest eyes. Some went to one place and some to another. But, of course, hearing about the trail isn't the same as walking the trail. Knowing where you want to go is the most important step in going anywhere. But in the end, if you really want to know, you have to go.

Isn't it interesting that our concept of learning comes from ancient concepts of sunlit meadows, of tracking previously made footprints, of following a trail, and of listening to the tales of those who have traveled before us?

Does it give us a better understanding of what we do when we learn about a new subject? Does it give us a better understanding of what we should be doing when we attempt to teach others?

## THE TEACHER
I'm full of emptiness now.
Nothing's left within.
I seek inside for a trace
Of anything that has been.

It's the time of day the students pass,
And I stop a passing friend.
I ask them about their plans
Now that school will end.

I hold him with my questions.
I can tell he wants to go.
But there are so many things
That he simply doesn't know.

It was many and many years ago
That I started on this path,
Proceeding without a clue,
Unaware of the aftermath.

He looks at his parting friends
And really wants to go.
It's plain he does not want to see
The things that I could show.

Now everyone has the left the hall.
There is loneliness in the room.
Light is filtered by the dust,
All is empty as the virgin womb.

I am left to reminisce
On how it all began.
How one thing led to another
Though I was unaware of any plan.

The child watched from the corner room
When he slammed the door, and she cried.
Then the child slipped from the house
And went to read outside.

In his books, the ships set sail. Arthur ruled benign.
Lawrence paced the desert sand.
And the diamond mines of Timbuktu
Are where he made his stand.

The books lined up in disarray and time
Piled recklessly on the floor.
As he escaped the time
Through books; the ultimate timeless door.

One last, lone soul hurries down the hall
But looks the other way.
I try to catch his eye to talk
To try and make him stay.

He hurries past, no friendly face,
No thought lives within.
He hurries to the future
Unaware of what has been.

Then alone, I recall yet other days.
Anger raging strong in tyranny,
When I would flee in fear and loss
Into eternal trees.

Guided by music of the Heavens,
The constellations in the sky,
An angel in the evening,
And a book that testifies.

I came to learn, to teach, and testify,
To change the world somehow.
But the world that I have come to
Holds no future for me now.

I crossed eternity to find no one here.
I came to spread the word.
But now I stand here all alone and
Discover that no one heard.

No monument is cast in bronze
In memory of the courageous dead.
No requiem was ever sung.
The last book has been read.

Dying men find no one to hear
For their lives do not remain,
Young men embrace their lives,
Their future, though unknown, is plain.

For this purpose, young men come
To find there's no one to fear.
In the end they cannot remain
And no one else can hear.

The ancient wisdom they have gained
By mortals not understood.
Carried with them beyond the grave
To build with God for good.

# 25

# THROUGH THE NARROWS

*"Well, perhaps; but I begin to think there*
*are better things than being comfortable."*
*George MacDonald*

I have a friend who is a river runner and knows a lot about water. I'm not exactly sure what a river runner is. Personally, I can't even walk on water on a still, small pond, but he can apparently run on the river. Anyway, when we went to dinner the other night, he told me that there was 10,000 cubic feet per second of water flowing in the Colorado River that day. They expected there to be as much as 35,000 feet per second flowing at the river's crest. He wondered how many elephants that would be equivalent to in weight.

I don't know too much about elephants, so I looked up what a cubic foot of water was. One cubic foot of water per second is 7.48 gallons, and a gallon weighs about eight pounds. A five-gallon bucket of water would weigh forty pounds. I happen to know that a five-pound bucket of honey weighs about sixty pounds, so I guess honey is thicker than water. But I have no idea how much a five-pound bucket of blood would weigh, so I can't really verify that blood is thicker than water.

Anyway, eight pounds' times 7.48 gallons of water would be 59.84 pounds of water in one cubic foot per second. So, 10,000 cubic feet of water, passing by a given point at any given second, would weigh 598,400 lbs. According to the Mozilla Firefox search engine, the African bush elephant weighs up to 16,500 pounds. I think I'll round that off at just 16,000, knowing how mammologists always exaggerate. If we divide 598,400 lbs. by 16,000, we get 37.4 standard elephants worth of water that passes a given point every second.

That's impressive! I think if you were running on the river, you would definitely want to stay on top because if you were on the bottom, things could get pretty heavy. Granted, the weight would be spread out over the surface of the river bed. So if you were to go under the water, it would be better to do it where the river is wide. That would provide an increased surface area for the weight to rest on. My friend explained to me that, unfortunately, that is seldom where people go under.

There is another problem here also. These calculations are only for the weight of elephants on top of you at one undefined point for one second. That means there would be 37.4 more elephants on top of you in just another second. I believe that could easily be called a stampede!

But how big is a point? I was about to ask, "What is the point?", but decided against it. Suppose you were to get in front of all the water coming down stream. I really don't see how you could get behind it very well, can you? You would then have the weight of all the water up stream pushing you downstream.

Let's suppose a point is six-feet long. That's how tall I am. You would have the first 37.4 elephants pushing down on you. Behind that there would be another 37.4 elephants pushing you downstream. Behind that would be another 37.4 elephants, and so on upstream. In just one hundred yards you would have the equivalent of 623 elephants, or 9,973,333 pounds, pushing you downstream.

I can see right away, even if my assumptions are a little off, that walking on the water isn't going to do anyone much good. If you are going to do anything, you had better run on that river, and run like heck. I guess that's why they call them river runners.

I asked my friend why he wanted to run in front of a stampede of elephants. He said it was fun and exciting. I understand that because I like to blow things up for the same reasons. I think humans are kind of like that. We probably could stay where it's nice and safe all the time, but we seldom do. I think even deciding to be born on this earth was pretty adventurous seeing as how a spirit in the pre-existence would have no idea whatsoever they were getting into. Maybe I'll take up rock climbing. It's really the only way to find out what's on top.

**THROUGH THE NARROWS**
Through the narrows there are few who pass.
The water is deep and the current strong.
Walls of rock, no blade of grass, and
Raging waters that rush headlong.

The water falls and the river roars.
If you enter, you'll be swept along.
No foothold for you on either shore.
To the river, you'll belong.

Day or night, the water runs cold.
Shadows light the deep crevasse.

Treacherous currents and hidden shoals
Fill the deep and dark morass.

From moment to moment, as you grow old,
Decision and action choose the way.
Any mistake may require your soul.
Be sure you want to pass this way.

Far below, beyond where the eye can see,
Around the bend that obscures the end,
Believing there you will be free.
If you stay afloat, you'll enter in.

The wisdom of the wise is so often unclean.
They'll tell you not to enter in.
But those who enter are not unseen,
And all who emerge are cleansed of sin.

Through the narrows there are few who go.
They hang in the balance one by one.
Once entered you can't escape the flow.
You cannot quit until you're done.

When the river runs out, and the narrows cave,
Then man will see his final days.
The fools on the bank, with their empty ways,
Will have no water to extinguish the blaze.

Through the narrows, to the sea,
The broad expanse of what will be.
At the end you will be free,
In sun and sand and serenity

# 26

# THUNDER ON THE MOUNTAIN

*"Past tears are present strength."*
*George MacDonald*

Don't you just hate it when your Mother was right? Well, it only bugs me when I didn't think she was right at the time, and then it turns out she was. My Mom always said, "Can't never done nothin'." She was actually too educated to talk like that. She just used that vernacular for effect. Boy did it affect me. It always made me mad! She would whip out those words when I was whining that I couldn't do something. I was probably whining because I didn't want to do something, but it was just as irritating.

What Mom was trying to tell me, of course, was that I could do anything I really wanted to do. I just had to decide to do it. Or maybe she meant that if one gives up, they never accomplish anything. If she had said it that way, maybe it would have been less irritating, but I wouldn't have remembered it so well. When we think we can't do something, we set our own limits. Not that we don't all have limits, of course. But boundaries are often self-imposed or artificial.

I was thinking about this while watching the supermoon recently. It's called a supermoon when a full moon, or a new moon, is seen at its closest proximity to the earth during its elliptical orbit. The combination of proximity and fullness makes it appear larger than at other times of the year. I thought that sounded kind of cool and wanted to see it.

My wife went to bed early that night, so, instead of following her, I went outside to see the moon. I sat on the tailgate of the truck for quite a while and looked at the night sky, the moon, and the Perseid meteor shower. Does that make me weird?

What made this whole experience unusual is that I realized I seldom look at the moon anymore. I used to be out, under the night sky, and playing games a lot as a kid. There was also night patrol in the military, long walks at night, camping . . . and often with my wife.

Anyway, the words from a James McMurtry song, "Levelland", came to mind.

"Mama used to roll her hair
Back before the central air
We'd sit outside and watch the stars at night

She'd tell me to make a wish
I'd wish we both could fly
Don't think she's seen the sky
Since we got the satellite dish"

What I noticed about the sky that night was that there weren't any borders. There were no edges that limited the panorama. It wasn't a movie, and it wasn't a television, iPad, or a cell phone screen. It was boundless. It was infinite. I had the feeling one gets when standing atop a mountain peak. It's the feeling I get when cresting the ridge west of Rabbit Valley and seeing that stark, yet beautiful, expanse of desert stretching as far as I can see into Utah.

Maybe I have spent too much time looking through a microscope, absorbed in those infinitesimaly, small worlds, bounded by a ring of darkness. Are there too many boundaries limiting my vision? Perhaps I don't see far enough anymore. It's like I can only enter the store through the entrance, and I must leave by the exit. I can only communicate the way the technology allows.

I am surrounded by the idea of "can't". You can't go there. You can't do it that way. You can't have that. Vision ends at the edge. I wonder what affect it has on a culture and a people when everything they see and do has boundaries. Could that be the difference between us and the pioneers? They saw vistas and night skies. We see pictures of vistas and photographs of night skies. My Mom might say that boundaries are just "can'ts", and "can't never done nothin'."

I think the deal about "can't" is the difference in "deciding to" and "feeling like". Sometimes, when I am having a really good time, I get sad thinking about how it has to end. That is strange because, when I am sad and having a bad time and people tell me that things will get better, I never believe them. In fact, I have been known to argue about it. "No! No! No! It will never get better. Never, I tell you!" Do you ever do that?

It is sad when a happy moment ends, but often that doesn't happen for a very long time. Even when the happy moment ends, it doesn't usually turn ugly or something. It just gets a little less happy. A long time later something sad may happen. This has been the pattern all my life, but I still feel sad after a certain amount of happiness.

I have an equally impressive history with bad things turning out okay, or even pretty good, eventually. They always do. So why don't I believe it when someone says it when things are bad? It's just as consistent. Am I just morbid or something?

Of course, I think a large part of the problem is that good and bad are relative terms, so it's hard to know where they fall on some kind of number line. Even worse, they are opposites which means they can only be understood in relation to one another. So if I feel slightly less good, I must describe it as feeling bad, or at least feeling worse.

So any change from one to the other, by any degree, must be interpreted as being good or bad. Logically, it follows that good and bad must go back and forth very dependably. Neither can last.

The last problem is that they are both feelings which are not even "things" anyway. You can't pick up a bad feeling and pet it to make it feel better. Good feelings run through your fingers like water. Actually that's just a metaphor because water is real, just insubstantial in the liquid state. Feelings are words we use for something going on in our minds somehow, someplace. You can't pinch a feeling and make it stop giggling.

I have this problem in Church sometimes too. We have a meeting every month where individuals can stand and express their feelings about the Savior and the Church. We call it sharing of testimonies. Well, sometimes I feel like bearing my testimony, and sometimes I don't. Because it is a little difficult to express our emotions on deep subjects in front of people, it is always easier to not do it.

I have decided that it is a lot better to simply decide that I am going to bear my testimony, and then, do it regardless of how I feel. After I've done it, I always feel good about it. So I think that doing things because you decide to is more reliable than waiting until you feel like it.

We need to learn that feelings are like night and day, winter and summer. They come and they go, and there really isn't much you can do about them. In dealing with them, it's far more reliable to just make decisions and carry them out.

## THUNDER ON THE MOUNTAIN

Moonlight blows through the garden
To the rustling of the leaves.
But there is thunder on the mountain,
And darkness on the face of the deep.
The Spirit moves 'cross the water.
Moonlight drifts my way.
But there's writing on the wall.
Can you tell? What does it say?

Candlelight flickers in the window,
A welcome where I am bound.
But there's thunder on the mountain
I can feel it in the ground.
Laughter that can tell us,
"I'm so glad you came".
But through my open window,
I smell the coming rain.

Evening perfumes hover.
There's a flash of colorful plume.
But there is thunder in the garden,
And a rock rolls from the tomb.
So in a happy stupor,
We wile away the day.
But there, just under the litter,
Is the world of putrid decay.

Stars shine in the Heavens.
Constellations guide the way.
But there's thunder on the horizon,
And lightning strikes the clay.
Music sings from the Heavens
Lending sweet dreams in my sleep.
Morning stars fall one by one,
Leaving darkness on the face of the deep.

# 27

# TIME

*"Work is not always required. There is such
a thing as sacred idleness."*
George MacDonald

Sometime in the 1700's, people began to conceive of time as linear in measurement. This was unlike earlier times when time was considered circular. There have been times when time has been perceived as random with no pattern at all. Most of the time, we say that time is uniform, but there are times when time does seem to go faster or slower.

Contrary to what many believe, I am not old enough to recall ancient and primitive conditions. However, the passage of time has always brought about unpredictable and dangerous changes, often resulting in dissolution and death. I have experienced my share of dissolution, and I can affirm it is dangerous.

For most of recorded history, though, humans have measured the passage of time in natural cycles that match the diurnal rotation of our planet, lunar months, and the rotating seasons. The measurement of cyclical time allowed greater organization of society and control over the elements. Humans learned to perform certain work, such as planting or hunting, at the "right times", or in the correct sequence.

Cyclical time introduced a moral dimension to mans' thought as well. Things could be said to have been done at the wrong or right time. In addition, each generation could begin to compare its behavior to that of its ancestor's behavior during similar periods of time. People who lived under a cyclical paradigm valued patience, relatedness of parts, ritual, relationships, nature, the healing power of time, and the symbols of resurrection and renewal.

Today, thinking of time as a linear experience, with a distinct beginning leading to a unique ending, is nearly universal in modern thought. The concept was recognized by ancient Greeks who hoped that reason would improve mankind's lot by enabling him to avoid mistakes of the past. The Romans felt that time led people down a path that culminated in a glorious destiny. However, it was the rise of the monotheistic religions that suggested that mankind's fate might be directional.

It was in the sixteenth century that the inventions of science, combined with the Reformation, began to spread through Europe leading people to speculate about the origins, and the end, of the earth. This linear timeline is carried forward in today's science where we debate the beginning of the universe and life, as well as the evolution the universe has undergone.

The idea of linear time assumes that mankind is on a trajectory of progress. Men disagree about when time began and when it might end. The overall assumption, however, is that we are on a straight line of progress towards better things. This thought so pervades modern America that it has shaped our entire culture. In contrast to cyclical time, modern culture value: haste, practicality, concentration, efficiency, analysis, direction, speed, power and control of nature.

Modern science seems to believe we are simply on an extrapolated timeline from the past. Perhaps, if one does not see any possibility of deviation from the trajectory in the future, they consider any deviations in the past insignificant. In fact, the past is assumed to have led to this moment when we are lucky enough to be in existence at the exact apogee of human existence.

One thing worries me. Straight lines do not always lead upward. They can just as easily be drawn downward. All measures of time that we use are based on measurement of repetitive activities such as the vibrations of an atom, the rotation of the earth on its axis, the lunar cycles, or the seasons in a year. Maybe those comparisons should be enough to give us pause about modern times. It's about time.

**TIME**
Time has both hands around my neck
If I want to make a difference yet.
I'm not living for myself anymore.
Seeds must die to be reborn,
And I must prepare for night to come.
Marching to the seasons drum,
In the morning, up with the dawn,
I can rest later on.

In the spring, I must sow my seed
If I expect to ever reap.
Before the leaves are upon the bough,
There are seeds to sort and ground to plow.

And calves to birth, sheep to shear,
Chicks to hatch, and brush to clear.
In the beginning, up with the dawn.
I can rest later on.

When the sun is shining high,
Perhaps then I can close my eyes,
And rest in shade from a leafy bough.
The sun is hot.  It's summer now.
Pray for rain, hoe the weeds,
Work till dark, there are mouths to feed.
In the morning, up with the dawn.
I can rest later on.

When the honey harvest is done,
Then I can rest in the setting sun.
Colored leaves decorate the bough.
The harvest's on, it's autumn now.
Pick the apples from the tree.
Harvest honey from the bee.
In the morning, up with the dawn.
I can rest later on.

When the leaves have fallen from the bough,
And before the spring ground to plow,
There's are saws to sharpen, fences to mend,
Seeds to order, and accounts to attend.
In the evening, I cannot rest my hand.
I must prepare to rest the land.
In the morning, up with the dawn,
I can rest later on.

# 28

# THE PAINTER AND THE PAINTING

*"Wherever there is anything to love, there*
*is beauty in some form."*
George MacDonald

Have you heard about the two, red-blood cells that loved in vein? Well, OK, but don't let me catch you repeating it then! The truth is I don't understand anything about the science of love. I think there are some subjects which just don't lend themselves to scientific explanations. We, scientists, need to recognize our limitations.

For example, I have never understood what it is that women find attractive in men. Honestly, what could any woman possibly see in a handsome, smiling, muscled, tanned, kid wearing tight jeans and driving a brand new 4X4 jacked-up pickup? He's probably no older than twenty-five, immature, and probably doesn't have a brain in his head. And girls, any guy with a great tan doesn't have a job! Then, what's with the sinister dudes? Why are women so surprised when they turn out to be, well, sinister? And don't even get me started on bass players in rock and roll bands . . .

If you girls really want true romance, you have overlooked a quiet group of substance, stability and culture. Well, OK, it's a unique form of culture, but it is one. I'm talking about science geeks, of course! There are so many advantages to dating science geeks that it just surpasses my understanding why women aren't more enamored with them.

In the first place, they are generally available. Lacking in social skills, motorcycles, and tight jeans, they have been overlooked for so long that there is an overabundance of them on today's market. Not only are they available, but another advantage in dating geeks is that other women seldom try to steal them. I can tell you from experience that my wife and I have been married for almost 45 years, and no woman has even made a pass at me. The only plausible explanation for this, of course, is that I am a science geek, and my wife has never allowed me to buy a motorcycle.

Geeks have other things going for them too. One is that parents almost always love them. They appear harmless, often make good money, and can fix things. That is no reason to marry someone, obviously. But it does remove some of the difficulties from life while you look over the passing parade of bass players.

Science geeks are surprisingly sensitive and romantic people once you get past their initial, social awkwardness. For example, you wouldn't want to miss out on Valentine endearments such as:

The Rosette Nebula is red.
The Pleiades star cluster is blue.
The universe is expanding
Like my love for you!

Note that it can be difficult to meet science geeks in the first place. They often have peculiar tastes in alternative music, so you seldom see them at concerts. They're even more rarely found in sports bars. They generally hang out in laboratories which have restrictive access, and they tend to socialize in groups where they discuss obscure and unintelligible topics. When seeking them out, you can be at a distinct disadvantage.

But here's a tip. Guys wear t-shirts with logos of their favorite bands and sports team, thus showing that they are sinister dudes or manly athletes, right? Well, science geeks tend to wear t-shirts with logos of software programs and science symbols emblazoned on them to show that they are, ah, well, geeks. Since there is a convivial rivalry about these things, you could try wearing one yourself. See if your tee strikes up any conversations!

Of course, the best way to meet science geeks is on the internet. Surfing the net allows science geeks to combine an activity with which they are comfortable – computing - with an activity they are uncomfortable with - socializing. Another strategy is to hang out in the junk food aisle of the grocery store.

Most importantly, science geeks thrive on mystery. So just keep being female, and they will be helplessly fascinated, with an emphasis on the "helpless" part. Beauty is in the eye of the beholder.

**THE PAINTER AND THE PAINTING**
I was born with the power buried deep inside.
God has said it is my fate.
From childhood on, it has been my guide.
The Prophets have labeled it "born to be great".

The King is the King, and you know what that means.
He wanted a painting made of his bride.
A King must provide the desire of the Queen.

A man with power has no place to hide.

I was summoned, then, to do the Kings will
And capture the image of his Queen.
To use the powers of my skill
To see in her what the King has seen.

    The King is the King, and you know what that means.
    Those who displease him are usually lost.
    A banquet was planned out on the green
    To unveil the canvas and learn the cost.

I was born with the power, buried deep inside,
To see the soul of all those I would paint.
I cannot paint lies. Believe me, I've tried.
I cannot make a devil look like a saint.

    The King is the King, and you know what that means.
    He was born with power on the outside.
    From truth and lie forever between,
    He cannot reveal. He must always hide.

I arrived with my canvas covered from sight,
And asked the King to grant my desire.
That when the painting was shown to the light,
He would bide time before expressing his ire.

    The King is the King, and you know what that means.
    He nodded and said he would grant my request.
    As noble and kind, he wished to be seen.
    Although in truth, that is surely a jest.

I displayed the canvas and revealed my fate.
The canvas was bare as the desert sand.
Then held up my hand for the King to wait,
And began to paint without his command.

    The King is the King, and you know what that means.
    But he had already stated his mind.
    So he waited to see how I painted the Queen

And to decide how I'd be consigned.

So much depends on the course that you take.
So I watched the Kings face, as I painted the Queen,
For any sign that I'd made a mistake.
And I changed my technique at the sight of his mien.

The King is the King, and you know what that means.
When the light dries and the colors fade,
He sees what he sees in his queen.
What I must paint is his shade.

So much relies on the course that you take,
But I was born with the power inside.
Both fool and the wise man may burn at the stake,
But my portrait is hung in the Kingdom with pride.

So what did I capture when I painted the Queen,
In the hour I spent to capture the scene?
Is it product or process that pleases a King?
The King is the King, and you know what that means.

# 29

# UNITED

*"Until a man has love, it is well he should have*
*fear. So long as there are wild beasts about, it is*
*better to be afraid than secure."*
George MacDonald

I was a soldier from 1966 until 1968. I left the military with mixed emotions: joy and gladness. Don't misunderstand me: I am proud that I served my country, and the experience was personally very beneficial. It's just that I am not really strong, courageous, or obedient. I think, by the end of two years, both the U.S. Army and I had come to grips with that. We parted ways on pleasant terms.

I mention this only because it illustrates an important point often missed by scientists. Since science is the study of the material world, scientists tend to be pathologically focused on material. The material world is made of physical objects. It is the nature of physical objects, so science tells us, that no two objects can occupy the same space at the same time.

Therefore, the study of matter is the study of diversity. Whether we are cataloging a new element, observing a specific force, or defining one animal from another, we are breaking the world up into smaller, separate segments. It stands to reason that, somewhere along the line we would divide people up into races, even though all humans have ninety-nine percent the same DNA.

By contrast, many important ideas and concepts are abstractions that can exist in the same space and time simultaneously. I experienced both joy and gladness upon discharge from the military. Every grandparent understands the sadness with which they watch a grandchild leave as well as the feelings of exhaustion and relief as they go.

In fact, some abstractions nearly always accompany each other in the same time and place. Young lovers are almost universally consumed with excitement and confusion simultaneously. I am assured that such emotions as love and irritation can occur simultaneously in wives. I have actually heard this from a number of sources although some are more trustworthy than others.

Another important limitation to science and the study of the material world is that the same object cannot exist in two different places at the same time. Interestingly, the same material object can occupy the same space at two different times. In fact, this latter idea is almost obligatory unless somebody moves the object in the interim.

Luckily for reality, spirituality, and romance, abstractions do not have these limitations. In fact, the situation is almost reversed. Two abstractions can exist at the same time in different spaces. How else could one explain romance than by acknowledging that two different bodies feeling attraction and affection towards each other simultaneously?

However, unlike the material world, it is not necessary for two abstractions to occupy the same space at different times. Abstractions, such as emotions, don't just sit there in the same place all the time. Love can be lost, or at least ,temporarily misplaced.

Anyway, I think all the attention being paid to diversity now days is the end result of the scientific revolution where everything has had to be categorized into separate physical spaces. So humans have invented races, cultures, political parties, religions and other separate abstractions as if they were material objects.

At the same time, it seems humans long for unity which can apparently only happen in the strange, spiritual world of abstractions. Only there can abstractions such as shared feelings and attitudes unite at the same time and space or occur at the same time in two different people. Perhaps this is one of the roles religion can play in our modern world if we don't try to see the abstraction as a material separation but as a unifying force.

Abstractions such as love, unity, commitment, and faith are the things that unite husband and wife. If these occupy the same people at the same time, there is unity.

### UNITED

On a rare summer night, with a second full moon,
The old man is awake while the earth slumbers.
Many beasts abound in the distant wood,
And he hears the echoes of their sounds.
Only they and he are out under lone stars,
But their cries cause him no alarm.
Just brothers in arms singing similar songs
Longing, pretending the past isn't gone

He often sits alone on a fallen log
Unable to sleep because of his weary mind.
He leans his back against a tree and hides,
Dark, in the shadow, from the silvery moon.
A sitting watchman, alongside a moonlit grave,
His mind alive with life gone by,
Of faces and hearts in new found love
And with memories of a recent past.

A past like so many others have had,
The bitter losses and memories of all mankind.
But their loss brings no solace to his soul.
Like them, he pretends his love is still alive.
If he could but raise the one beneath the snow.
If he could only howl in the pitiless night.
If he could chase, and rend, and tear,
With anger, as his brothers might, this night

In reverie his spirit drifts just above the snow,
When, startled, he thought he heard her speak.
He felt her presence in the gloom.
He felt her spirit in the night.
She stood before him, reaching for his hand,
With the old light shining yet in her eyes.
While his brothers called her from afar
And while he wondered at the sight.

Then from her mouth he heard her words,
In the soothing voice he knew so well,
And she said "Fear not, we'll be together soon.
A love like ours cannot be denied.
God has declared salvations plan
And eternal love for all mankind."
Then her image began to fade, and he sat alone
In the sparkling moonlit snow.

In desperation, he clutched for her reaching hand
Even as the dawn began its eternal glow.
His brothers stood around in a ring and howled out
Their farewell in the snow.

The two faded together above the grave.
With lightened heart, he bade the sun farewell.
No longer a partial human soul,
At last, again, he became one whole.

# 30

# WANDERERS IN A STRANGE LAND

*"With wandering eyes and aimless zeal, . . . "*
*George MacDonald*

Every moment the world begins anew.  Nothing remains the same. The new children born change the world they are born into.  The old people who die change the world they leave behind.  The child that is aborted didn't get to make its contribution.  The person whose death was postponed by medical personnel will behave differently with their remaining time because of their close encounter with mortality.

Scientists discover new laws and ways of controlling the world, but they never received your permission to do so.  There are secret minds, turning hidden wheels, that will change your life forever.  Without your knowledge, people in high places make decisions that affect you none-the-less.

A person breaks a marriage covenant, and the resulting heartbreak and destruction in a family create economic hardships that ripple through your community.  Personal decisions made in desperate situations change the course of events for thousands, if not millions, of individuals.

The world begins anew and is forever a new, strange place from what it was previously.  There are but two choices.  Proceed boldly into the future, or sit quietly and safely in your ever-changing present until you are overwhelmed.  Each moment we are all strangers entering a strange new world.

**WANDERERS IN A STRANGE LAND**
They say there is a land of eastern mists,
From which the sun rises for its morning tryst.
A land of dense forests and wild inhabitants;
Beautiful women who are strong and vibrant,
And savage men with wild, shaved heads.
A land of water from which the sun has fled,
But one can never be sure.

They say there are men who think it best
To lead their families into wilderness.

The wilderness of trouble and asylum,
To mystic lands of vapor and revelation.
Adam, Abraham, Nephi, each into chaos;
What seemed at the time hardship and loss
Was, for them, salvation.

The ages reveal what cannot be seen.
The fate of man is revealed scene by scene
By how he spends the moments of his time.
The will of God, his ancient paradigm,
Is revealed in seconds that are as a thousand years.
Men have left their homes in tears and jeers,
But it is those who remain who perish.

Go boldly, then, though you cannot be sure
If your love of God is strong and pure.
Like ages past, you will be led.
And a nation born after you're dead.
So go forth with faith and eagerness.
God's children live in the wilderness.
May God bless the "wanderers in a strange land."

# -31-
## SEVEN BLACK CROWS

*"How strange this fear of death is! We are
never frightened at a sunset."*
George MacDonald

It's comforting to know when the end of a year comes. It's always after 365 days. You can prepare for something like that. It's predictable.

I could tell that my students anticipated the end of my lectures. I wasn't as precise as the clock or the calendar, but I hardly ever went over time. The students always started packing up long before I was through talking. I found it strangely comforting as it was a relief to know that I was about done.

Pretty much everything comes to an end. As far as I can tell, it isn't the endings that bother humans. It's their difficulty in predicting the endings. That's why the calendar is cool. You know when the year will end, and it's over in an instant.

We all die. Of course, on the other days, we all live. Not a bad trade-off really. So far I have lived over 25,000 days. I am being purposefully vague here, but I only have one day of dying still to come.

I don't like the idea of bucket lists. Having one would make me feel all rushed to get things done. I would be especially ticked off if I got everything done and then found out I had to sit around for years waiting to die. My personal goal is to take things nice and easy up to the day I die and then kind of taper off.

It would be convenient if we knew when we needed to have our affairs in order. Of course, some of us would just put the job off to the last minute anyway.

So why, if all living things are made of the same basic stuff, do all have different life spans? How come the tortoise gets 73,000 days and humans get, on average, a third of that? Maybe I need to slow down. . .

It turns out that the difference in lifetimes is because of the number of heartbeats we are allotted. Way back in the 1930's, a Swiss named Max Kleiber studied the relationship between mass and metabolism. He came up with a formula predicting the energy burned per unit of weight is proportional to an animal's mass raised to the three-quarters power. In plain words, the smaller you are, the more calories it takes per ounce to keep you alive.

However, if you eat a lot of food, you have to metabolize it quickly so that you will have room for the "more food" you are going to eat. Metabolizing food requires energy and generates heat. Getting rid of heat and circulating energy require heartbeats.

It turns out that, at least for mammals, there is an allotted number of heartbeats per lifetime: about a billion. A relaxed shrew, which may be an oxymoron, has a resting heartrate of about 850 beats per minute. At that rate they have a life expectancy of about two years. Some whales have heart rates of ten to fifteen beats per minute. This buys them about two hundred years.

I figure that, on the average, my heart rate has been about seventy beats per minute for most of my life. Of course, there were those occasions when I did stupid, terrifying things that increased the rate. There is also my wife who still takes my breath away. I figure those times are offset by all the hours at work, during which time, I didn't really do much of anything. Seventy is probably a good average.

So, if you calculate this out to a billion heartbeats, I'm already dead. And I have been since I was about twenty-seven years old. Humans seem to beat the billion heartbeat rule with brains. It's not that we live longer because we live smarter, but it's because we use medicine and interventions to extend our lifespan. I have to admit though, It's not very predictable.

## SEVEN BLACK CROWS

There, in the hollow of the hills, I see
Seven black crows waiting for me.
Low light on the rim and the wind informs
In the distance of a gathering storm.

Shadows long of waving grass,
Clouds appear and then pass,
Respite from the sunlit glare
In the darkening electric air.

The silver crescent shows low in the sky
As the sun begins to die.
I fear the crows crowding my sight.
I also fear the coming night.

I no longer see the crows on the plain.
Only rhythms of unrest remain.
Night and day still revolve.
Crows and ghosts all resolve.

Nothing to lose because it's already lost.
I can see the coming holocaust.
Within the cloak of darkest night,
Do the crows remain, or have they taken flight?

Black horizon against black sky,
Watched by a million twinkling eyes.
Serenade of night-time sounds
As I traverse the crow's playground.

Seven black hills and I walk slow.
Uneven ground, unknown foe.
Never before have I been here,
And it's the seventh hill I fear.

As I stand at the last hill's base,
The dark seems to have lifted just a trace.
All those in the dark watching me
Seem as if they can no longer see.

With great fear, I reach the height,
And look to see in the fading night
What lies ahead and waits for me
On the other side, what will be.

There, In the hollow of the hills, I see
There are no crows awaiting me.
Low light on the rim, a gathering light,
And my love stands there dressed in white.

# -32-
## VISIONS

*Anything large enough for a wish to light upon,*
*is large enough to hang a prayer upon.*
*George MacDonald*

I often have these brilliant ideas just before I go to sleep at night. I try to write them down, so I won't forget them by morning. In the mornings, I sometimes have long lists of ideas and experiments that are usually written in an unintelligible scrawl. They are great ideas though! The few that I can decipher are so brilliant I can only mourn the loss of the others. Anyway, I have been saving these ideas on three by five cards.

Note cards are old fashioned, but computers hadn't been invented when I first started. By the time I had access to computers, my list was so long it was too daunting to transfer into some kind of searchable data base. Unfortunately, computers and data bases weren't either of my early ideas.

My intent was to eventually enlist student-slave labor to tackle these issues. So I quietly kept them to myself while lesser scientists struggled to find answers to minor questions like the structure of DNA. Of course, DNA had been doing just fine on its own for about a million years.

It now seems obvious that the list is so long I will never get around to probing a fraction of my ideas. Therefore, I have decided to share some of the more pressing issues with my readers in hopes that others might pick up the gauntlet.

Probably one of the more urgent questions is, "Does gender play a role in lie detection?" This is a complex issue, since it involves not only knowing which gender tells the most lies, but which gender is best at telling and detecting lies. As a matter of fact, it's even complicated to tell which gender one is studying anymore. This is significant to determine if one gender has an inherited affinity for detecting lies of the opposite gender. We attempted a pilot study a few years back, but the results were unreliable because we forgot to exclude lawyers and politicians from the study.

A second, urgent study would involve the effects of dehydration on Karaoke singers in bars. Alcohol is a known desiccant, and singing undoubtedly increases loss of moisture through respiration. So I am told the question that remains to be answered is whether or not drunks sing longer than sober patrons, thereby monopolizing the microphone. This research has been difficult for me to deal with since I don't go to bars or sing karaoke. The second problem, I am told, is that there are no sober patrons singing karaoke in bars.

Triboluminesence would be a really cool name for a rock band. That was an original note to myself on my idea pad. However, later I thought it might also be the answer to our worlds energy needs far into the future. See, sometimes, if you break chemical bonds just right, that action will release energy. You can demonstrate this by crushing Wint-O-Green mints in the dark. If done correctly, you can see sparks as you break the sugar-crystal bonds.

Numerous other hard candies need to be tested. Adequate dental insurance would be required too, I'd guess. Scientific research can get so complicated at times. Anyway, could this source of energy be harnessed in some way to supply our energy needs for the future?

I haven't pursued this project yet, because I can't find any place that's really dark anymore. Think about how light our woprld has become between street lights, glowing clock faces, and LEDs on electronic equipment. I suspect further research will require going to a remote wilderness area. Hmm, that just might add to the attractiveness of this project. Well, as long as there is a Holiday Inn close by.

Another note from my night stand said, "Thermophelgistantion". I think. I can't quite make it out. Something like that. If you can make any sense of it drop me an e-mail, will you?

Finally, I noted several years ago that when I stopped working in mosquito control, I gained weight. More recently, as I have taken up beekeeping, I seem to have lost some weight. This suggests a strange correlation between human obesity and insects that has not been adequately explored. I don't think it is dietary since I seldom eat insects . . . voluntarily. As usual, further research is needed.

### VISIONS
My night-time dreams never survive,
But are lost in the space of absurdity.
They are the visions of which I'm deprived.
They are the visions I never quite see.

Day dreams are the ones that I finally believe
Because they stay in my memory.

They are the ones I tried to achieve,
With faith that somehow they really might be.

They are the dreams that have carried on
In face of apathy and willful ignorance.
Never quite there, but never quite gone,
Searching and seeking in reverence.

I once believed that they were dead,
Perhaps killed by their very own sword.
But love is a binding of multiple threads.
Love is patient until it grows bored.

There is a reason that dreams come at night,
And that daydreams hold us close.
Because we see more clearly without our sight
With the aid of Father, Son, and Holy Ghost.

I dreamed I came from a previous sphere.
I dreamed I only came here to grow.
I dreamed I knew why I am here,
And that I have someplace else to go.

Night dreams persist twisting and turning,
Somewhere between joy and sorrow.
Daydreams persist and keep on burning.
They're the best way into tomorrow.

# DESERT DREAMS

*To have what we want is riches; but to be able to
do without is power.*
*George MacDonald*

Have you ever stared at something until it starts to look like something else? I mean, when you haven't been drinking. I haven't, but I am told some people can. It must be true because there are numerous people who can look at something and see it completely differently from the way I do.

If you're really good at seeing things differently, you might consider a career in topology. This is a field of mathematics that studies geometric problems that depend, not on the exact shape of the object, but on the way the object is put together. In other words, topologists study the properties that are preserved in an object when it is deformed, twisted, or stretched. Tearing, however, is not allowed.

In topology, a circle is considered equivalent to an ellipse because it can be deformed into that by stretching. A sphere is topologically equivalent to an ellipsoid. Well, I don't know about you, but I just flat disagree with that. I don't know who makes up these rules. Well, actually I do know. Topologists make up the rules.

As usual with mathematicians, I don't know what they are talking about. Topologists do have some useful ideas though. For example, a topologist might have a hard time determining a doughnut from a coffee cup. See, if one were to twist and poke a pliable doughnut into the right shape, the hole could become the handle and one could poke a cavity to become the cup portion. It's the same structure, just twisted differently.

Now, before you get all excited, I have already patented the idea. This is going to be huge - a doughnut with the coffee poured into it! No more messy dipping. We'll start with a simple, glazed doughnut, but there should be no reason we can't expand quickly into cinnamon and cherry jelly.

Anyway, that gives you a feel for how important topology can be in solving scientific problems. Another major contribution of topology is the famous "hairy ball theorem". This theorem was first stated by Henri Poincare, a French mathematician in the 19th century. He stated that, "there is no non-vanishing continuous tangent vector field on even dimensional n-spheres."

That's what I'm told anyway. I think it's French. Translated, I think it means that if you try to comb a hairy ball flat, there will always be a cowlick. It took me years of experimentation with comb-overs to prove the validity of this now-accepted theorem.

I had a topological experience once, and now I have a greater appreciation for this field of endeavor. It turns out that when a retina becomes detached, the retina being the field of nerve endings in your eye, someone either has to solder it back in place with lasers or put a rubber band around your eye ball to push it back to the retina. It's sort of a "If the mountain won't go to Mohammad, Mohammad will go to the mountain" kind of situation. Wait, can I say that kind of thing?

Anyway, I ended up with a cool rubber band that distorts my left eyeball. Now, when I close my right eye and look at a round clock on the wall, it looks oblong. Well, it used to anyway. Apparently one's brain can somehow correct for these kinds of things because now the clock looks round again. I'm back to normal, but it's not as entertaining as when I could switch back and forth looking at the clock with each eye.

This raises the interesting question about whether topology is even real or not. I mean, if we stare at something and deform it in our minds, will our minds just un-deform it after a while? This is a hopeful thought. Perhaps someday, the brains of all those people who see things differently from the way I see them will eventually un-deform, and they will finally begin to agree with me.

TOPOGRAPHY

Topology is not the same as topography. Topography is the arrangement of the natural and artificial physical features of an area, the lay of the land. Of course, it also includes the nature of the climate, whether it be hot or cold, dry or wet.

Topography can be just as complex as topology. It's just not restricted to doughnuts and bald heads. I guess you can change topography, a little like topology, except it takes a lot more work. Then it's never clear whether the new topography will be any better. Topologists never have to worry about whether it is better or not because it's all imaginary anyway.

I had a topographic experience once. I was raised where topography is critical. In fact, my whole life has been a topographic experience. Having spent most of my life in Colorado, it's hard to have anything else. It's nice to ride your bike downhill, but there is always the realization, in the back of your mind, that you have to ride up again sooner or later.

Everything in Colorado is either upstream or downstream. It's pretty important what goes on upstream because whatever it is will be coming my way. It is comforting to Coloradoans to know that both Texas and California are downstream.

In reality, I live on the very edge of Colorado. East of me is the entire range of the Rocky Mountains. But on the south and west it's high desert plains that eventually shade and blend down into southern Utah and Arizona. I especially love the Navajo nation and Monument Valley. It's red buttes and towering mesas are somehow spiritual to me.

**DESERT DREAMS**

There are no fairies in the desert
For there is no place to hide.
There are no goblins in dry valleys.
Without water, they've all died.
There are no banshees on the mesa,
No dwarves or Elvin kings,
No giants or hobgoblins,
Only wind and brittle things.

But there are imaginations,
Echoes of mankind's sin.
Watching clouds will slow your run.
Do not whistle, you'll call the wind.
Skin walkers are the desert witch.
Chindi, shadows from the sky.
Frogs come from sacred rain.
Evil comes when it is dry.

The desert is both clean and dusty.
I walk where it is hot and cold,
Avoiding the wash which once was dry.
The sudden flood renews the old
Like the desert and its purpose.
Be it evil or be it good,
Neither condition can exist
Unless the other could.

Talking God, Changing Woman,
Corn People, or the ant,
I need it all in my quest:
Sunlight, animal, and plant.
The beauty of the morning
Is as the beauty of twilight.
The sun God of mid-day
Becomes a million suns at night.

I go in beauty across the sand
Surrounded by good and ill.
With harmony in my heart
I go in beauty, peace be still.
I walk with purpose, avoid the dark,
Avoid the crest of the hill,
In harmony with the world,
Seeking for all that's truly real.

# CHAKO CANYON

*"When I can no more stir my soul to move, and
life is but the ashes of a fire; when I can but
remember that my heart once used to live and
love, long and aspire- O, be thou then the first,
the one thou art; be thou the calling, before all
answering love, and in me wake hope,
fear, boundless desire."*
*George MacDonald*

The first requirement for being able to do science is time when you have nothing much to do. Free time eventually creates in the human mind a feeling of boredom, and you know what they say, "An idle mind is the Devil's playground."

Anyway, one of the reasons for the falling off of science, technology, engineering and math scores, STEM, is that young people have too much to do. By the way, you know the old adage "i before e except after c"? Well, that doesn't apply to science. As I was saying, most good science came out of not having anything else to do, sometimes called necessity.

A case in point is astronomy. Astronomy was probably the first science developed, and it all took place because there really wasn't anything else to do. Try to imagine what it must have been like three thousand years ago: no i-pads, no electric lights, no books, no late night TV, no smart phones, and no rooftops. There would have been nothing to do but pay attention to what was happening at the moment or play guitar.

I suspect one of the first things people noticed was that the sun came up every day and went down every night, just like clockwork. They might have phrased their thoughts differently since there weren't any clocks, but the regularity must have been impressive in what seemed a chaotic world.

With nothing to do at night but watch the stars, it became apparent to them that certain stars stayed in the same place, or changed their position, with regularity. Sometimes the regularity was through the night, but other regular movements occurred over longer cycles. With only a couple of years of observations, an attention span that seems beyond imagination in our world, they began to see relationships between the positions of the sun, moon and planets to the seasons of the year.

These simple observations allowed humans the ability to make weather forecasts. After three thousand years, you'd think we'd have gotten better. They also helped people control their lives better by enabling them to decide when to plant crops, shop for winter clothes for school, or plan vacations.

Another difficulty man had in some parts of the ancient world, was the terrain. As hard as it is to imagine, many people in the world have no horizon. Well, I suppose they all have one. Most are just boring, made up of trees and the actual skyline. There are no mountains, or other interesting landmarks around to mark the movements of the sun.

This required them to construct structures, along which they could site between the sun or moon and some other point on a man-made horizon, to determine the extremes of the annual cycle. That's where things like Stonehenge and Native-American Medicine Wheels came in.

In the Grand Valley of Colorado, we have our own built-in Medicine Wheel called the Grand Mesa and Mount Garfield. On December 21, of any given year, the sun rises in the southernmost part of the sky. For us that will appear to be somewhere around Delta, Colorado.

If you get up early, you will have nothing to do. You can stand at a given point in your back yard and determine a landmark on the south side of Grand Mesa where the sun rises on the horizon to mark the winter solstice.

You will have to wait until June 21 of each year to find the point on the horizon to mark the longest day of the year, or the summer solstice. At this time, the sun will rise somewhere north of Mt. Garfield, off towards Meeker some place.

I stand at the north corner of my home and site over the beehive to mark the winter solstice. Then I stand at my homes south corner and sight over the native bee nest-post to mark the summer solstice. If you can't find time, you can just Google it . . .

**CHAKO CANYON**
There is no question more important than
When the sun will come again.
Nothing has so preoccupied men
As the urgent question concerning when
The sun will finally come again.

128

Time, often measured with earth and rock,
Laid out in array like a clock.
See cairns that mark the horizons view?
When the sun is there, time starts anew.
Then we know the past is through.

Sun daggers and moon shadows mark the time
Of moon standstills and the summers prime.
Long nights for long years to mark the signs
And symbols found on ancient shrines
To mark when again the sun would shine.

These ancient men with eternal vision
Understood patterns of the Heavens.
They watched, waited and remembered when
Last things were as they were then,
And by so doing blessed all men.

But we are all no wiser still
Than those ancient men on ancient hills.
In fact, we may know less than they
Who used rock cairns to mark the clay,
Marking that special, holy day.

We do not know when the light will come.
Blinded by our artificial sun,
We think it is always light.
Continual progress, always bright,
Unprepared for winters night.

There is no question more important than
When the Son will come again.
Nothing has so preoccupied men.
As the urgent question concerning when
The Son will finally come again.

# 35

# OUT OF CHAOS

*No words can express how much the world owes to sorrow. Most of the Psalms were born in the wilderness. Most of the Epistles were written in a prison. The greatest thoughts of the greatest thinkers have all passed through fire. The greatest poets have "learned in suffering what they taught in song." In bonds Bunyan lived the allegory that he afterwards wrote, and we may thank Bedford Jail for the Pilgrim's Progress. Take comfort, afflicted Christian! When God is about to make pre-eminent use of a person, He put them in the fire.*
*George MacDonald*

I have decided to get in shape. And the shape I have chosen is a triangle... On second thought, all those sines and cosines would probably confuse me. I have enough trouble following street signs. But the shape of things, including signs and sines, have always interested me. Why do things even have shapes? I mean, a chair has a shape suited to sitting in. But why do things like rocks, trees, rivers, and crickets have the shapes they have? And who gets to decide what shape they will be? No one ever asked me. As you can see, I also have problems with tangents . . .

But to have a shape, something must be a solid. It's hard to have a shape if you can't hold it, and only solids can hold their shape. Solids are a result of the interface between order and disorder, and the arrangements of elemental particles called atoms.

Atoms strike me as odd things. I think of them as particles, but I am told they are mostly empty space with a few smaller particles like electrons and protons floating around. These packages of mostly space can be packaged together in different ways to make what we call the "three states of matter".

The nature of what physical state we perceive is less about which specific atoms are involved, although that is often important, and more about how close together these packets of space are packaged. Atoms, which are mostly empty space, when packed close together become the thing we call a solid. And solid things have shapes.

Someone has said that "solids are those parts of the physical world which support when sat on, which hurt when kicked, and kill when shot." So if I understand this correctly, when we pack something that is mostly empty space closely enough together, we get a solid.

But, of course, the space in atoms isn't really empty. It is just empty of material. Uhm, what else is there? Well, I am told that the space in atoms is filled with such things as electronic fields. Fields are empty space, so you see the space inside atoms is filled with fields. Is this getting clearer?

Electronic fields can actually fill space in the same way that a magnetic field can fill space. If one take two magnets and brings like poles together, you will feel a resistance filling the space between the two magnets. Depending on how strong the magnets are, and how strong you are, it may be very difficult, or impossible, to push the two together. The space between the two magnets seems to be full of something.

It doesn't really matter which atoms we are talking about, just how close together they are, for us to experience solidarity. Wait, isn't that a political movement? Anyway, a solid is a substance in which atoms and their accompanying fields are packed together very closely. If the atoms are not closely packed, they can slide around and across each other much like two magnets with like poles seem to. Such a substance can't hold a shape and is called a liquid.

Much of what we experience in the physical world, the shape of things, depends simply on how closely together the atoms are packaged. Solids are closely-packed empty spaces, liquids are less closely-packed empty spaces, and gasses are empty spaces packed loosely into a larger empty space. Seems perfectly clear to me.

Of course, once atoms are brought into close proximity to one another, they have to fit together according to their shapes, like patterns on wall paper. That is where it becomes important which shape of atom is involved. Some fit together in hexagons, some as cubes, and some even as triangles. That's my kind of shape.

## OUT OF CHAOS

From out of chaos, crystals came.
With flashing moons the atoms rained.
Random oscillation atomic correlations,
Chemical imagination, order is born!

From strange attractors, patterns form.
Unpredictability becomes the norm.

Parts extraneous fit instantaneous,
Though extemporaneous, a fractal forms!

From out chaos, crystals formed.
Null and void, a universe is born.
Creations premise, from parts miscellaneous,
Mundane and precious, Adam's world is born!

Out of nothing, but solar flares
The order of God was declared.
And, in the vast firmament,
Order came, and nothingness was rent!

From matter unorganized,
Without form or name,
From out of chaos, a crystal came!

# 36

# SEED

*The truly wise talk little about religion and are not*
*given to taking sides on doctrinal issues. When*
*they hear people advocating or opposing the claims*
*of this or that party in the church, they turn away*
*with a smile such as men yield to the talk of*
*children. They have no time, they would say, for*
*that kind of thing. They have enough to do in trying*
*to faithfully practice what is beyond dispute.*
George MacDonald

I watched the flock of birds sweep across the sky. Suddenly they veered left in unison, as suddenly darted to the right, then swooped downward toward the ground, only to rise upward and circle back again. Their movements were perfectly coordinated and graceful. There were no collisions. The flock, as a whole, seemed more graceful than any single bird, even better than a well-choreographed dance troop.

The only unusual thing about this sight was that it was all taking place on a computer screen, and the birds looked more like little, grey, paper airplanes than birds. This was because I was watching a computer animation, not a real flock of birds. Yet, I could see no difference in their behaviors from the real thing.

Most of us assume that birds play a game of "follow the leader". We think the bird in front leads, and the others follow. But, apparently, this isn't so. The computer simulation was established by a set of simple rules instructing each "bird" in the animated flock to hold its position steady against all birds around it. That is, the "bird" on its left should be kept at a fixed distance and a specific angle. If another "bird" shows up on the right, the distance and angle should be maintained as well.

There is no "leader bird". Each bird in the flock simply follows, or reacts to, the movements of other birds nearby. Orderly flock interactions arise from these local, bird-to-bird interactions. Much of human society seems to develop in the same way. Neighborhoods seem to function for the common good in many cases, especially during emergencies, and long before large organizations can have much effect.

This is an example of what scientists call self-organization. Self-organization occurs when groups of autonomous particles interact in such a way as to give rise to organized patterns or behaviors. Birds are not the only animals to self-organize. Ant colonies, termites, beehives and slime molds all self-organize. The immune system operates without centralized direction. The development of the human embryo occurs without central control.

Central to the idea of self-organization is the concept of decentralization. Many social organizations appear to occur in a decentralized manner. Traffic jams appear to develop spontaneously. Market economies develop complex behaviors and patterns that arise without leadership. Adam Smith argued against centralized control of economies more than two hundred years ago. One of the unique contributions of American political thought was the idea of states' rights and decentralized national government.

Decentralization appears to occur on certain scales. When things get too large, it becomes increasingly difficult to manage all the details, and efforts to do so often create, paradoxically, disorganization. For example, a beehive may hold over one hundred thousand individuals. The queen does not tell the hive what to do. But upon appropriate cues, the hive grows a new queen and splits into two. It's interesting that the Soviet Union and IBM decentralized their management within one day of each other.

The common assumption is that, when something seems complex, it must have a complex explanation. However, time after time, that doesn't appear to be true. At one time in the past, people thought that the gene must be a protein because proteins were the largest, most-complex molecules. We assumed that only the most complex molecules could account for the amazing diversity in the living world.

Instead, we have discovered that inheritance and diversity are explained by a chemical code made up of just four elements. These elements behave much like a nested binary code with two elements dictating a small set of choices. The other two possibilities determine the final message. There is no boss in the cell!

Time after time, complex things are shown to be constructed of simple things utilized in unique ways. Organization can occur, indeed does occur, without central direction. It's almost as if decentralization is a part of the plan.

## SEEDS

Seeds grow, though sometimes they're slow.
In the sun or under the snow.
A little water, a little sunshine,
Some grow in the ground.
Some grow in my mind.

Seeds know what they need to know.
What color to be and what shape to grow.
Ancient knowledge preserved through time,
Some grow in the ground.
Some grow in my mind.

Seeds go wherever they go.
Some on the wind others go with the flow.
They don't really know what they will find.
Some grow in the ground.
Some grow in my mind.

Seeds above and seeds below,
Right and left wherever we go,
Seeds in front and seeds behind,
Some grow in the ground.
Some grow in my mind.

Some are wild, some grow in a row,
Some are plain, and others for show.
Pieces of life completely refined,
Some grow in the ground.
Some grow in my mind.

Some are long, some seeds are round.
Grow the stem up and grow the root down.
How they should grow has been predesigned.
Some grow in the ground.
Some grow in my mind.

Some seeds are fast. and others are slow.
Some grow high. and others grow low.
Seeds grow in the sunshine.

Some grow in the ground.
Some grow in my mind.

# 37

# SISTERS OF THE WOLFEN MOON

*It is by loving and by being loved that one can*
*come nearest to the soul of another.*
*George Macdonald*

We don't usually think in terms of a "volume of light". We tend to speak of brightness. But the amount of anything available in a given space is called a volume. "Please pass a quart of light," sounds strange, doesn't it?

But think of it this way. If you focus light with a lens on a given point, it will form a cone. Anyone who has ever eaten an ice cream cone knows that a cone has a volume and that one can only increase the volume by increasing the height of the ice cream. The diameter is fixed by the physical diameter of the opening to the cone.

So if you want to know exactly how much an ice cream cone will hold, you calculate the volume of a cone with the formula: volume = 1/3(Area of Base)(Height), or 1/3 (pi x r2) (h). Remember that from middle school? Me neither. I had to look it up.

Don't worry about it much, though. No one makes the pointy ice cream cones anymore, which messes everything up. But, when you focus light it still makes a cone, so you can still calculate the volume of light in your cone. Knowing this, we could if we wanted, calculate how much light we receive during a given time period in volume instead of brightness.

Did you know that if you have 5000 cubic inches of light for several hours, you can cook food with a cone-shaped solar cooker. If you make the base wider or the height taller, you will get more light and hence more heat. But increased volume also takes longer to heat up so it is a trade off.

The advantage of using a cone is that, if there is a reflective surface on it like tinfoil or something, it tends to focus the light into the tip to give a higher temperature at the focal point for better cooking.

Of course, If you are more interested in ice cream, you probably shouldn't put it in this kind of cone. With ice cream it is probably easier to change the height of the ice cream than the diameter. I'll have three scoops of ice cream, please. But for my baked potatoes, please pass eighty seven quarts of light.

## SISTERS OF THE WOLFEN MOON

Once there was a summer girl with winter-colored skin.
She had a darling sister, a darker shade of twin.
Friends were they, and sisters true.  They lived in northern woods.
They swore no one could have their hearts.  That no man ever could.

Then they awoke one morning with leaves falling all around.
Snow hovered in the sky, and frost perched upon the ground.
The geese they were a-flying, silver angels drifted down,
And by the night of the quarter moon, snow covered the ground.

They locked the door, built a fire, and pulled their blankets to their chin.
They kept the raging winter out and bolted their wandering in.
And when at last the snow did cease and the sky filled up with stars,
The moon shown through the window and spied Winter from afar.

So tapping at her window with icy pebbles, he did call
The summer maid with winter skin to dance in his snowy halls.
What awoke the darker twin when her sister did take flight?
Was it cold from the open door or a howling in the night?

She ran into the moonlit night and followed footsteps in the snow.
Then standing at the forest edge, she saw Winter in the meadow.
She was waltzing with the moon. It seemed that she could fly.
And as her sister stood and watched, Winter waltzed into the sky.

The cruel moon had taken Winter.  And the darker twin, in despair,
Cried out in savage fury.  Her sorrow filled the air.
There were others in the forest who heard her mournful cry.
Through the shadows they came to her with movements softer than a sigh.

A band of the broken hearted, they joined in her cries.
They cried their anger at the moon for its promises and lies.
Among them was a dark one who lost his brother to the stars.
So, united in their sorrow, she gave to him her heart.

The summer girl with winter skin did not return again.
And leaving with the dark one, neither did her twin.
And so the sisters parted, parted far too soon.
But the dark twin calls her sister each year at the Wolfen moon.

# 38

# FLAMING MOTH

*A beast does not know that he is a beast, and the*
*nearer a man gets to being a beast, the less he knows it.*
*George MacDonald*

They say diamonds are a girl's best friend. I gave my wife a beautiful beetle once. It was from Brazil, and I had embedded it in plastic. I think she would have preferred a diamond. She apparently didn't know, or maybe she didn't care, that beetles and diamonds have something in common. However, diamonds have nothing in common with plastic. Maybe that was the problem.

But ask Lauren Richey of Springville, Utah about beetles and diamonds. For her, beetles have turned into gold. Or am I mixing my metaphors here? It all started with that annual spring right called the Science Fair. Richey started doing science fairs in junior high. At first she was not terribly successful, but then she began to win regional and national awards for research projects, usually involving light and photonics.

As a senior in High School, Richey read a paper suggesting that iridescent butterflies might contain photonic crystals. She admits that she read the article mostly because of the beautiful blue butterfly on the front that stirred her interest. She approached John Gardner, a professor in the BYU Physics Department, with her idea to examine a beetle, *Lamprocyphus augustus,* a shimmery green beetle, with an electron microscope. And sure enough, the beetle exoskeleton contained structure similar to photonic crystals.

Excuse me, but when I first heard this story, I didn't know what a photonic crystal was. Apparently these structures, which are relatively rare, resemble the arrangement of carbon atoms in a diamond crystal. This crystalline shape can affect the propagation of electromagnetic waves in the same way that a semiconductor can in a computer. This would allow a computer to operate on light waves instead of electricity, a coveted goal in computer science. While the crystals in a beetle are too fragile for use in a computer, they might serve as a template for the manufacture of more sturdy structures.

But the significance of this story isn't really about photonic crystals. It is about a student who did college-level research while a senior in

High School and published a scientific paper as a freshman in college. Since then, Lauren, a sophomore in physics at BYU, has published two more research articles in professional journals and taken over 13,000 electron micrographs of beetles. She is presently examining beetles that contain photonic crystals of an opal-like nature rather than those like a diamond.

The Intel International Science and Engineering Fair, Intel ISEF, is the world's largest international pre-college science competition. Individual schools run the series of local competitions. The science fair may be one of the best kept secrets among middle school and high school students. Students can receive many different awards, in numerous categories, as well as get a taste of scientific research. Many students receive cash prizes or scholarships. In some cases, like Lauren's, participation can set the stage for an entire career.

Anyway, I was sure to show my wife this story since I think she never properly appreciated the beetle I gave her. She's probably sorry now that she knows all about photonic crystals. She also never wore the mosquito ear rings, or the silver cockroach pendant I gave her. Wait. Silver cockroaches! That gives me an idea.

## FLAMING MOTH
He's a plain brown wrapper,
But there isn't any shame.
No need for color in the moonlight.
He tumbles in before the rain,
Slipping and sliding he came,
Keeping out of sight.
> In some ways he's like a man of the cloth,
> No one even noticed he came.
> I know you think you should bet on the moth,
> But I think I'll bet on the flame.
Searching for an evening flower,
Her scent is in the air.
Hiding on the bark of a tree and
Moving gently with great care.
Fluttering from here to there,
Twisting in the evening breeze.
> In some ways he's like a man of the cloth.
> He acts as if this were a game.
> I know you think you should bet on the moth,

But I think I'll bet on the flame.
Keeping one eye on the silver orb,
The light directs his flight.
Sending feelers out into the dark,
Alert for a night-long fight.
Hampered by many-faceted sight,
His path an earth-curved arc.

    In some ways he's like a man of the cloth,
    He can't help it, he's not to blame.
    I know you think you should bet on the moth,
    But I think I'll bet on the flame.

He follows the distant and ancient glow.
It marks eternal flight.
It has served him for a million years.
Yet the closest star is the most bright,
A blazing flame in the dark night.
Though that way ends in tears.

    In some ways he's like a man of the cloth,
    Blinded by the light of the flame.
    I know you think you should bet on the moth,
    But I think I'll bet on the flame.

# 39

# DUST

*"The lightning and thunder, they go as they come:*
*But the stars and the stillness are always at home."*
*George MacDonald*

I don't know what the school board was thinking when they planned for school to start August 12th this year. That is the peak night for the Perseid Meteor Shower. The students will have homework and have to be up early the next day to catch their buses. I guess it's more important to sit in class and do worksheets than experience the real world.

Listen kids, just stay up late anyway and watch the Perseid Meteor Shower. Then sleep in the next morning. If anyone gives you a bad time, just tell them I said it was OK. Of course that won't lessen the consequences you will face one little bit, but you can claim it was all my fault.

How important can a meteor shower be? The Perseid shower generally produces more falling stars than any other. As many as 100 meteors can be seen in an hour during the Perseids.

There are other meteor showers, to be sure. There is the Orionids in October, the Quadrantids in January, the Lyrids in April, and the Eta Aquarids in May. Notice all of these are during the school year. The Perseids offer the last opportunity for school kids to stay up all night watching falling stars while eating junk food.

My experience is that the best education is guerrilla education. Nothing is so thrilling as to learn something just because it's fun and has absolutely nothing to do with the curriculum and high stakes testing. Usually guerrilla education requires some collaboration and inventiveness. I suggest you enlist younger children. Anyone old enough to think of a guerrilla educational scheme probably can't find enough "mature" people to actually help carry it out.

Obtain a fairly large aluminum or plastic container. Don't bother your teachers with this because they have paperwork to do. The container should have a wide mouth and hold a considerable amount of water. Fill this container with several inches of water and place it in a secure, but high, position such as on top of the roof. Don't bother your parents with this as they will question the necessity of getting on the roof and say it is too dangerous. Make sure there is open sky above with no obstructions above the container.

Leave the container there for several days. Three or four weeks is ideal, but keep the water level up. While the Perseids peak this year on August 12 and 13, 2013, they are active to a lesser degree over several days before and after.

When a meteor enters our atmosphere it ignites from friction with atmospheric particles. This burning becomes the falling star. As the meteor is consumed it emits clouds of dust particles which settle over the earth. Some of that dust will land in the water traps you have provided. Because of its origin, meteor dust contains a large amount of iron compared to earth dust.

Next obtain a plastic bag. Don't bother your mother about this. She will want to know why you want it and will feel compelled to explain how much it costs. You will also need a good magnet. If you do not have one in your possession, you can probably "borrow" one from school. Don't bother your teacher about this as he will explain that it is for school use only, and he can't let you take it home. You might learn something unauthorized. Just be sure to return it.

Place the magnet in the plastic bag and swirl it around in the dust-trap water. Any iron dust particles in the water will adhere to the plastic bag, attracted by the magnet. These are meteorite dust particles. Now take the bag out of the water. Remove the magnet and place the plastic bag in a bowl of fresh water. The iron particles will dislodge and fall into the water.

You have just collected stardust. The particles can be four billion years old and from as far away as Pluto. You can show them to your parents, teachers, and school administrators now. However, don't bother them until after school. You don't want to disrupt the educational process.

**THE DIG**

There are dusty shapes floating in the sun
But when I try to see them they come undone
I guess that's why I'm restless trying to find
The things I see floating in my mind

The truth of dust pours out the door
And I almost hear footsteps across the floor

As if I haven't chased dust in the past
Only to find she moves too fast

From dust to dust for dust thou art
The dust fills and chokes my heart
And as the dust disappears
I can no longer see you here

Once upon a time this world was green
But now it's dead and dry and clean
Dig, dust, and scrape the stone
For dusty shapes and forgotten bones

Shadows lie upon this tomb
And makes shapes floating in the gloom
In the past I hope to see
The things only visible when I sleep

Dusty shapes in the shifting sand
Not quite clear what lies upon the land
I guess that's why I'm restless trying to find
The things I see floating in my mind

# 40

# KINGS OLD

*"the punishment of the wrong-doer makes no*
*atonement for the wrong done."*
*George MacDonald*

"O say can you see, by the dawn's early light,
What so proudly we hail'd at the twilight's last gleaming,
Whose broad stripes and bright stars through the perilous fight
O'er the ramparts we watch'd were so gallantly streaming?
And the rocket's red glare, the bombs bursting in air,
Gave proof through the night that our flag was still there. . . ."

The 4th of July is our day to celebrate freedom, although we aren't free to set off the rockets anymore. We can still get up early by the dawns early light if we want. But let's face it, the only reason we see the twilights last streaming is because we're waiting until its dark enough to see the bombs bursting in air. Does anyone sit outside on a summer night anymore?

I think the 4th is the last really-male holiday. That's probably sexist, huh? But you have to admit it's the only holiday where it's legit to blow things up and act all militant and aggressive. Of course, you have to be properly licensed to actually blow stuff up now days. But you can watch other people do it. What a thrill.

One of the fundamental concepts of physics is that as things heat up, the molecules move about faster, push away from each other, and therefore expand the space they occupy. I suppose one could take advantage of this concept to make a rocket, theoretically, of course. Let me illustrate this principle of gas expansion, due to heat, with a theoretical model. I'll just make it a tiny model for simplicity's sake. "Theoretically" means this is all just a theory. It helps to understand a principle if one knows how it might be used practically.

Suppose that you took a little paper match, although theoretically you could use a wooden kitchen match as well, and laid it on a square piece of tinfoil two times the size of the match. The head of the match should be on laid in the center of the tin foil. About half of the match stem should extend beyond the tinfoil.

Then lay a pin or needle on each side of the match with one end even with the match head and the other end extending beyond the tinfoil. Once these are aligned, fold the tin foil tightly over the match head and pins from the bottom and then from the sides. Be careful to pinch between the match and the pins. This is so that when, I mean if, you remove the pins you will leave behind, strictly theoretically of course, two tiny exhaust ports for expanding gases.

Theoretically, one needs a safe launch site to launch rockets. Just like NASA, it would probably be best to utilize a space with a lot of bare ground and a safe landing site. Propping the match up at an angle would usually ensure that the match acts like a rocket and not a jet propelled car. Of course, this won't be necessary because these instructions are just illustrative of the principle of heat expansion anyway.

Now, if something like a separate match or other heating device were to be used to ignite the match head inside the tin foil, the interior burning match head would heat the enclosed gases. The gases would expand. While some of the gas would escape out the exhaust ports, the internal pressure would theoretically propel the match stick itself out of the tin foil and launch it as a tiny rocket.

Of course, no one would actually ever do this. However, it does explain, on a small scale, how the professionals create and fire off rockets. Only some macho male, of which there are very few left, would ever even think of actually trying to make one of these, so don't worry about any explosions or anything. I hope you enjoy the rockets' red glare this year.

## KINGS OLD
There is nothing new under the sun.
Many men have walked this road indeed.
Many Kings think that they have won,
Until on their death bed they lie and bleed.

With stone and lance and grey goose wing
Our fathers fought for bloody years,
And wrenched tyranny from the kings
And laid their lives on funeral biers.

So they bought us freedom at mortal cost,
Opposed the God anointed creed.

And now we watch as their gain is lost
And whether king or queen, we know the breed.

There is no lesson for us to learn.
It matters not what they may claim.
We know tyranny comes if kings return.
A queen or king under any name.

"Where are you going?  From whence you came?
Give me tribute from all your wares!
I will protect you in freedoms name!
I will use the guns you may not bear."

Kings bring division and hosts of spies.
Queens sell justice, deny, and delay.
While money breeds like carrion flies
As a king of old like a donkey brays.

There is nothing new under the sun.
You need not guess.  Nothing is hid.
As done before it shall be done.
Just the same as kings old did.

Forgotten bondage of heart and brain.
In ancient days when knights were bold
Did our fathers die to be bound again?
Suffer not queen or king as of old.

# ABOUT THE AUTHOR

Gary McCallister is an American author, musician and scientist. He is Professor Emeritus from Colorado Mesa University where he taught and did research for over forty years. He is the author of numerous scientific articles, books on a myriad of topics including music and religion, and the award winning author of a popular weekly science column. He has also produced over twelve music CD's and is a luthier specializing in hand crafting mountain dulcimers. He is a member of the Church of Jesus Christ of Latter Day Saints. He is married and the father of four children and seventeen grandchildren, so far. He has long been influenced by the Christian writing of George Macdonald, C. S. Lewis and J. R. R. Tolkien.

# ABOUT GEORGE MACDONALD

George MacDonald was a Scottish author, poet, and Christian minister. He was a pioneering figure in the field of fantasy literature and the mentor of fellow writer C. S. Lewis.

## OTHER BOOKS BY GARY MCCALLISTER

**MUSIC**

Making More than Music 2014

First Songs With the Mountain Dulcimer: history, instrument, and simple songs 2015

Hymns on Mountain Dulcimer: Learn to play the mountain dulcimer using hymns 2016

**SCIENCE**

Hanging Out With GRAVITY: Galileo's gravity game 2015

Seriously Silly Science: A science reader for the whole year – and some of it is even true 2015

A Convenient Truce: A cease fire in the war between religion and science 2016

**NOVELS**

Walking Man 2015

All available on Amazon.com